PERENNIALS FOR THE
PACIFIC NORTHWEST

PERENNIALS FOR THE PACIFIC NORTHWEST

500 Best Plants for Flower Gardens

MARTY WINGATE

Photographs by Jacqueline Koch

SASQUATCH BOOKS
SEATTLE

. . . .

To the memory of my mother, LaVerne, who started me on the garden path

. . . .

Printed in China
Published by Sasquatch Books
19 18 17 16 15 14 13   9 8 7 6 5 4 3 2 1
Editor: Gary Luke
Project editor: Michelle Hope Anderson
Cover and interior design: Anna Goldstein
Interior composition: Sarah Plein
All cover and interior photographs: Jacqueline Koch
except pages 25, 82, 83, 100, 102, 122, 139, 161, 201, 202, 226, 260: Blooming Nursery, Inc
page 238: Jerzy Opioła

Library of Congress Cataloging-in-Publication Data is available.
ISBN: 978-1-57061-893-2
Sasquatch Books
1904 Third Avenue, Suite 710
Seattle, WA 98101
(206) 467-4300
www.sasquatchbooks.com
custserv@sasquatchbooks.com

CONTENTS

INTRODUCTION

First, there were flowers. Not first in the grand scheme of the world, but certainly first in the hearts of gardeners. As young gardeners, we are enchanted with zinnias and marigolds, spears of salvia and mounds of impatiens. They last only one summer, but that satisfies us—until we discover that there are other flowers that don't have to be planted out year after year. These flowers appear on their own every year, not just once, but repeated—perennially. As our gardening years increase in number, we may look to even more permanent members of the garden: trees and shrubs. But our hearts never abandon perennials; the affection runs too deep.

Here in the Pacific Northwest, we can grow a wide variety of perennials; in fact, our plant spectrum is limited mostly by a lack of heat in the summer. The English style of perennial border has been embraced wholeheartedly in the Pacific Northwest, and why not? Our climates are similar enough that we, too, can grow just about anything we want (we say boastfully).

The flower border as a one- or two-season event, however, has been altered to fit our own style. In the Northwest, the garden is a year-round show that relies heavily on perennials, with strong support from trees and shrubs. Whether our opportunity to garden year-round has affected the variety of perennials available or vice versa, Pacific Northwest gardeners can now find plants that offer interesting foliage, form, or flower from late autumn through winter.

Flowers are possible in winter: hardy cyclamen (*Cyclamen coum*) dot the floor of the winter garden with pink, and the Christmas rose (*Helleborus niger*) opens its pure white flowers in December. But we also have the mounding form of the evergreen hardy geranium (*Geranium macrorrhizum*), rosettes of silver-leaved rose campion (*Lychnis coronaria*), and seedheads of purple coneflower (*Echinacea purpurea*) to decorate the dark months. The winter landscape may be more subtle, but it offers its own delights.

So if we can grow almost anything, and there are thousands of asters, hardy geraniums, and daylilies from which to choose, where do you begin in order to learn about and find the best perennials for your garden? Your daunting homework assignment is done for you in these pages. It may sound like drudgery to cross-check several reference books and plant availability guides, then to check with local sources and experts but, well, someone had to do it.

The purpose of this book is to provide descriptions and growing information for as many perennials and perennial-like plants for Northwest gardens as possible. Our accommodating growing environment can make choosing perennials seem like an overwhelming task, but our mild climate does have its limitations, especially when it comes to how much heat a plant might need to grow well. National and international perennial books are useful to a point, but don't take us into our own Pacific Northwest gardens.

Here, you'll find out how to choose and grow perennials as well as learn design ideas and tips. The extensive A–Z list of perennials provides you with plant descriptions and suggestions for use in the garden. Whether you don't know a marigold

golden marguerite
(*Anthemis* 'Susanna Mitchell')

from a *Monarda*, or you want to add another *Epimedium* to your growing collections, here you'll find plants and information you can use.

Gardeners needn't think that the perennials they see on the shelf at the big-box stores are the only game in town. This book shows you that there's more to perennials—much more. In addition to having the information here, in your hand, for choosing and taking care of perennials, it's sometimes good to get up close and personal with these versatile plants. Fortunately, you don't have to go far. There is a wealth of garden inspiration in our own backyards, so to speak.

For instance, hardy plant organizations throughout the Pacific Northwest (see Local Plant Societies, below) have an active agenda of meetings, symposia, and workshops all about perennials. The Hardy Plant Study Weekend is a lively and well-attended event that rotates each year among Northwest cities including Portland, Eugene, Seattle, Vancouver, and Victoria, BC. International speakers make these meetings worthwhile, and the private garden tours arranged for the events are always a hit. The Washington and Oregon groups also organize Open Garden days, when members' gardens are available for touring.

This is based on the highly successful Open Garden scheme in England. Your garden doesn't have to be perfect to be included on the tour (otherwise, no one's garden would be on the tour).

LOCAL PLANT SOCIETIES

Oregon

Hardy Plant Society of Oregon, 828 NW 19th Avenue, Portland, OR 97209-1504; 503-224-5718; www.hardyplantsociety.org. Membership includes a monthly newsletter, a quarterly journal, the Open Garden Book, and use of the lending library.

Willamette Valley Hardy Plant Group, P.O. Box 5942, Eugene, OR 97405; 541-344-0896; www.thehardyplantgroup.org, info@thehardyplantgroup.org. Eugene is one of the regular venues for the annual Hardy Plant Study Weekend, which rotates around the Northwest.

Avid Gardeners, P.O. Box 50808, Eugene, OR 97405; www.avidgardeners.org. A Eugene-based group of both professional and amateur gardeners who gather regularly for lectures and other events.

Salem Hardy Plant, P.O. Box 2027, Salem, OR 97308-2027; www.salemhardyplantsociety.org, shps@earthlink.net. Activities of this regional organization include monthly meetings, plant sales, and an Open Garden program.

Washington

Hardy Plant Society of Washington, P.O. Box 77556, Seattle, WA 98177; www.hardyplantsocietywa.org. An educational organization offering lectures, workshops, and field trips.

Northwest Perennial Alliance, 8522 46th Street NW, Seattle, WA 98335; 425-647-6004; www.northwestperennialalliance.org. NPA is responsible for the planting and maintenance of the mixed border at the Bellevue Botanical Garden. Meetings are held throughout the year with top-notch speakers; small study groups meet in various locations; and there's a monthly newsletter.

Whatcom Horticultural Society, P.O. Box 4443, Bellingham, WA 98227; 360-738-6833; www.whatcomhortsociety.org. The Whatcom society offers excellent lectures and symposia and encourages members to open their gardens during the year. The quarterly publication The Social Gardener contains articles and news about horticultural events.

British Columbia

Vancouver Hardy Plant Society, 1122 Deep Cove Road, North Vancouver, BC V7G 1S3; www.vancouverhardyplant.org. This BC group is one of the rotating hosts of the Hardy Plant Study Weekend.

Victoria Horticultural Society, Box 5081, Station B, Victoria, BC V8R 6N3; www.vichortsociety.org. The Victoria group holds regular meetings and is a host of the Hardy Plant Study Weekend.

Another source of gardening information is the Great Plant Picks program, initiated in 2001. This consortium of professional gardeners, nursery people, and general hort-heads from throughout the maritime Pacific Northwest recommend the best trees, shrubs, perennials, vines, and bulbs for our gardens. To that end, each year they produce a list of their picks for great plants. Find out more about GPP at www.greatplantpicks.org.

Public gardens like those listed here teach us about plants and how to use them. At any time of year, gardeners can be found, cameras and notebooks in hand, at these displays that provide learning experiences as well as moments of gardening contemplation. Regular visits to public gardens are a delightful necessity.

And of course, many fine nurseries also show gardeners how the plants they sell can grow in a garden. Knowledgeable nursery professionals give great advice on your plant choices.

PERENNIAL DISPLAYS

Visit these gardens and nurseries for inspiration and education when it comes to perennial gardening.

Oregon

Dancing Oaks Nursery, 17900 Priem Road, Monmouth, OR 97361; 503-838-6058; www.dancing oaks.com. Retail nursery in the country with vast yet intimate garden displays and a place to picnic.

Joy Creek Nursery, 20300 NW Watson Road, Scappoose, OR 97056; 503-543-7474; www.joy creek.com. Retail nursery with ideas for all about just where and how to grow their plants.

Northwest Garden Nursery, 86813 Central Rd., Eugene, OR 97402; www.northwestgardennursery. com. A wholesale nursery with extensive mixed borders and displays for shade, sun, dry, moist. Gardens open by appointment.

Washington

The Northwest Perennial Alliance border at the **Bellevue Botanical Garden**, 12001 Main Street, Bellevue, WA 98005; www.bellevuebotanical.org. Open daily from dawn to dusk; no charge. Planted on a slope and encompassing all manner of soil type and exposure. Come any month of the year to see what's in bloom, in seed, just coming up, or dying off beautifully.

PowellsWood Garden, 430 S. Dash Point Road, Federal Way, WA. 98003; 253-529-1620; www .powellswood.org. A three-acre, English-style garden open by appointment.

The Soest Herbaceous Display Garden, Center for Urban Horticulture, University of Washington Botanic Gardens, 3501 NE 41st Street, Seattle, WA 98105; 206-543-8616; www.depts.washington.edu/ uwbg. Planted in 1998, the Soest garden display beds incorporate various soil types and sun exposures. Strong on design despite its educational bent, the garden is attractive in all seasons. The Elisabeth C. Miller Library—filled with gardening books, magazines, and more—is only steps away.

British Columbia

University of British Columbia Botanical Garden, 6804 SW Marine Drive, Vancouver, BC V6T 1Z4; 604-822-3928; www.ubcbotanicalgarden.org. Besides an impressive arboretum, the gardens include good displays using perennials.

VanDusen Botanical Garden, 5251 Oak Street, Vancouver, BC V6M 4H1; 604-257-8335; www.vandusengarden.org. This 55-acre garden in the middle of Vancouver has a decidedly British feel to it—certainly due in part to the large and impressive European beeches (*Fagus sylvatica*) on the property. Set against this are various garden displays that appeal to any style, incorporating perennials in natural and formal settings.

Now that you have some ideas for where to go for garden inspiration, delve into these chapters for some step-by-step instructions in using perennials in your garden. Chapter One, Getting Started with Perennials, covers the basics on choosing, buying, and planting. Chapter Two, Designing with Perennials, explores techniques for designing your garden. Chapter Three, Maintaining the Perennial Garden, details the gardener's tasks: editing the garden, mulching, saving seeds, and dividing. In Chapter Four, Perennials from A to Z, you'll find the details on cultivation and garden uses for a comprehensive plant listing that includes not only a wide range of perennials, but also a few exceptional bulbs and a smattering of ornamental grasses. Only the best for you. At the back of the book, there's an index of common plant names. Now get growing!

goatsbeard (*Aruncus aethusifolius*)

Chapter 1

GETTING STARTED WITH PERENNIALS

WHAT IS A PERENNIAL?

One of the most common questions from new gardeners is, "What is the difference between a perennial and an annual?" So let's define the difference immediately. An annual is a plant that grows, flowers, sets seed, and dies in one year. Sweet peas are annuals, as are zinnias, cosmos, and sunflowers.

A perennial is a plant that lives for at least three years in the garden. Many perennials are herbaceous—they die back to the ground in winter, leaving not a trace of their existence. Other perennials leave behind a rosette of leaves close to the ground, in anticipation of the next growing season. And a few perennials flower during the winter.

Some plants that we treat as annuals, such as petunias, have been bred from plants that are perennials, or perhaps even shrubs, in their native environment—usually one much warmer year-round than our Pacific Northwest. So is a petunia an annual? The answer begins to get a little fuzzy, and not to beginning gardeners alone. It is usually enough to know that in our climate we must treat it as an annual because of our growing conditions.

lily-of-the-valley (Convallaria majalis)

Between annuals and perennials are biennials. These are plants that live two years. The first year they grow, and the second year they flower, set seed, and die. Biennials are also a cloudy issue to gardeners. Foxgloves, some verbascum, and parsley are biennials, but gardeners know that those plants may not necessarily die the second year. And so now we hedge our bets by calling verbascum a "short-lived perennial," which could mean that it lives two, three, or more years. But when it does die early, you can't complain, because you were warned.

And then there are shrubby perennials and perennial-like shrubs. Shrubs have persistent woody stems aboveground. Garden penstemon have persistent, somewhat woody stems aboveground, but are considered perennials, not shrubs. Lavender's woody branches are a giveaway to its classification—it's a shrub, but because it's small and flowers profusely, you may find them in the perennial section of the nursery. Plants such as lavender and sun roses (*Helianthemum*) are often called subshrubs.

Definitions set forth limits and boundaries, but it seems that definitions, like rules, are made to be broken—or at least stretched a bit. That's why what you read in a book or magazine seems to run contrary to the basic definition of "perennial." Bulbs, too, are often separated from a list of perennials, because their underground storage systems are actually modified leaf buds and not the fibrous roots we see when we dig up a plant such as a daisy.

While this book sticks with the definition of perennial as a plant that lives for at least three years, it seems natural to include a few plants that extend the horticultural definition yet fit so easily into a garden. And so this book includes a couple of bulbs (such as *Allium*) and biennials (such as common foxgloves, *Digitalis purpurea*), the occasional fern, and a well-placed grass or two.

CHOOSING PERENNIALS FOR YOUR GARDEN

There's no denying that most of us fall in love with a plant, buy it, and then try to find a place for it in our gardens. Although we never actually outgrow this practice (and why should we?), eventually we begin to look at choosing plants by matching their characteristics to our garden.

It's fortunate that most of our gardens include so many different microclimates and soil conditions that we can find a place for a broad range of plants. The rest of the world may believe that the Pacific Northwest is always gray and rainy, that we are the world's largest shade garden, but we know better. We do have a summer, which is dry, and we can grow more than just woodland plants. The contrasts can be dramatic: The sun's

For gardeners, a plant name consists of the binomial (the two-word scientific name) for the species, such as *Campanula persicifolia* for peach-leaved bellflower. The two italicized words together are the species, similar to your whole name including your first and last name. The first word is the genus, a large collection of plants; together with the second word, it makes up the species. There may be many kinds of sea holly (*Eryngium*), but only one yucca-leaved sea holly (*Eryngium yuccafolium*). Often, gardeners use the comman name for a plant—evening primrose for *Oenothera*—and sometimes the most common name is the same as the botanical name (*Clematis*). Here, you will find it both ways.

Some plants are classified as a subset of the species, because they have noticeably different characteristics. These plants have three words in their name, separated by "subsp." for "subspecies" or "var." for "variety." Nursery tags may or may not include these designations.

The word "cultivar" is a combination of the words "cultivated variety." A cultivar has been bred on purpose or selected from a group of seedlings or was found as a genetic mutation in nature. Sometimes the characteristic for which a cultivar was chosen and named comes true from seed, but most often these plants need to be asexually propagated—by division, tissue culture, or root or stem cuttings (see Chapter Three, page 35). The cultivar is written in single quotation marks, such as *Heuchera* 'Green Spice.'

Although a plant's scientific name is governed by strict rules, the cultivar name is up to the person doing the naming. Cultivars are often named for a person ('Ann Folkard'), a place ('Langtrees'), or the cultivar's most noticeable characteristic ('Orange Flame'). You may see the same cultivar name for entirely different plants: there is a *Brunnera* 'Langtrees' and a *Dicentra* 'Langtrees'. Sometimes namers get downright poetic ('Sops in Wine'). Guidelines say that cultivar names should not sound like Latin, so they aren't confused with the binomial, but this rule hasn't always been in effect; so you'll see many names in single quotation marks, such as 'Roseum' or 'Plenum.'

In catalogs and on plant tags, plant names often get mixed up, changed, misspelled, confused, and misapplied. Sometimes one or two of the words are left out. It's like playing telephone: What was whispered in the first person's ear ("play the record loud") mutates almost beyond recognition by the time it gets to the last person ("buy stock now"). In this book, every effort has been made to list the correct binomial and cultivar of these perennials, but that is not to say that you won't find them in a different form when you get to the nursery.

Where a plant has been reclassified (that is, taxonomists have realized that it was put into the wrong group), you'll find the alternate plant name in parentheses with the abbreviation "syn.," meaning this is a synonym. At the nursery, you may find either name.

umbrella plant (*Darmera peltata*)

effect is accentuated up against a south-facing concrete wall, while the north side of a building can be in deep shade for much of the year.

Combine those varying light characteristics with the wide range of soil types, and we can grow just about anything we want—even within one garden. Even a suburban-size lot can include patches of heavy clay and gravelly loam. Hot and sunny with sandy soil? How about a stand of the architecturally cool *Euphorbia characias*? Heavy soil in the shade? Brighten it up with golden creeping jenny (*Lysimachia nummularia* 'Aurea').

A garden should be a symphony of movements that lasts all year. Sometimes that means choosing plants that have more than one ornamental characteristic. At other times it means choosing a variety of plants that each have but one moment of glory, but at different times. This creates a succession of interest in flowers and foliage.

In the great gardens of England, the long, deep herbaceous borders may lie dormant in winter, looking like a barren wasteland. That isn't a concern, however, because the borders are often discreetly hidden from the house by yew hedges and old brick walls; those structures take over when there are no flowers. Out of sight, out of mind. But at Great Dixter in East Sussex, for example, where the long border previously was known mostly for its midsummer glory, pains have been taken to extend the border's show to begin in early spring and last well into fall.

In the mild maritime Pacific Northwest, it isn't difficult to find plants and combinations that carry through winter. The sections below provide a snapshot of one section of a garden; in a whole garden, the effect can be multiplied or mixed. (For more on designing with perennials, see Chapter Two.)

For Year-round Interest

Let's take one small piece of ground and see how it can be planted with four perennials for year-round ornamental characteristics.

- *Bergenia* 'Winterglut'
- *Penstemon* 'Garnet'
- *Carex testacea*
- *Helenium* 'Moerheim Beauty'

Three of these four plants have showy flowers. The *Bergenia* has deep red stems topped with deep pink flowers in late winter. Penstemon flowers in June and again in September if cut back after its first flowering. Helen's flower (*Helenium*) begins blooming in late July and can be cut back for more flowers in autumn.

But what this combination has is staying power. Whether the plants are in or out of bloom, there is always something to see. Of the three, *Helenium* is the only herbaceous plant. The deep reddish tones of the *Bergenia* and the orange grass *Carex testacea* stand out in the winter landscape, and the green stems of the penstemon (before being cut back themselves in late winter) add a bright touch.

For a Succession of Blooms

Now let's take that same small piece of ground and look at planting for a succession of blooms.

- *Pulmonaria* 'Roy Davidson'
- *Omphalodes* 'Starry Eyes'
- *Hosta* 'Blue Cadet'
- *Cyclamen hederifolium*
- *Aster divaricatus*
- *Helleborus* x *hybridus*—a dark one and a white one

This collection of shade plants lends a cheery note throughout the year. In light shade, lungworts (*Pulmonaria*), also known as soldiers-and-sailors for their two-tone blooms, start flowering in early spring. The flowers open pink and turn blue (although there are pure white cultivars, such as 'Sissinghurst White'); they are set off by green leaves with silver spots. Following close on the lungwort's heels are the intense blue flowers of *Omphalodes*, a low-growing charmer.

The blue-green foliage of the small *Hosta* 'Blue Cadet' has already unfurled and carries interest through the summer when its spikes of lavender flowers appear. In late summer, the bright pink

---

Understanding Hardiness Zones

The U.S. Department of Agriculture (USDA) has established eleven hardiness zones that are based on the average minimum winter temperature. The USDA hardiness zones are often listed inclusively (zones 3–8), to show not only the minimum temperature that the plant will survive, but also to suggest the plant's upper temperature limit. Much of the Pacific Northwest is zone 7 or 8.

For many of the plants listed in Chapter Four, you'll find that the upper zone limit is listed in national reference books as zone 7. That would certainly cut out a great deal of our region. The national guides do this because USDA zone 8 also includes much of the South, where hot, humid summers are the death of many delicate perennials. What we share with the South is an average minimum winter temperature—but our summers are a world apart. And apparently we are the only ones who know this.

toadflax (*Linaria purpurea* 'Canon Went')

flowers of the hardy cyclamen appear, followed by its marbled foliage, which stays through winter. Also in late summer, the wood aster (*Aster divaricatus*) blooms in a mass of tiny white starry daisies. And by late December, the hellebores are sending stems of flowers up.

BUYING PERENNIALS

Doing some research before you buy a plant goes a long way toward arranging plants in your garden for maximum effect with minimum problems. But often a nursery visit results in on-the-spot purchases, so you also need to be able to read and interpret the plant tag. These tags can provide you with loads of information about the plant, including its name (botanical and common—see Understanding Plant Names, page 4), hardiness zones (see sidebar, page 7), most suitable aspect (see Planting Perennials, page 13), bloom time, and ultimate size.

What plant tags don't tell you is exactly how the plant will grow in your garden. Your garden is different from everyone else's, and the effects of every condition whether large or small—from how much sun and water to what kind of micronutrients are present in the soil—can change the way a plant looks. The flower color may be slightly different, the plant may be more hearty (and hardy), or it may languish. That is something you learn in time, and it makes your garden a kind of scientific experimental ground from which you and others can gather anecdotal information. There are no control groups; you are not able to know the exact conditions under which a plant grows in someone else's garden.

That's why there is such lively discussion among gardeners: Our gardens provide us with a never-ending source of entertaining anecdotes. "My *Eupatorium* gets taller than that." "My hosta 'June' grows in full sun and looks great; I don't see why people say they need shade." "I tried growing rocket and it died." "I planted rocket and now it has spread all over my garden."

Where to Buy Perennials

Just as a sign that says "Antiques and Collectibles" can cause some people to step on the brakes, the word "nursery" on a sign gives gardeners pause. Why not stop and take a look? It's like window-shopping at Nordstrom: We don't have to buy anything; we just want to see what they have. And maybe have lunch or a coffee. And maybe get that one gorgeous blue poppy that we haven't seen anywhere else.

A gardener's favorite local nursery is the source of most of the plants in that person's garden, but it doesn't stop there. The world is full of purveyors of plants for our shopping pleasure.

I Can't Find That Plant Anywhere

Don't you just hate it when you read about a really cool plant, and then you can't find it? The plant may be rare, or it may be so popular it's difficult to track down, but either way, our resources for finding plants are growing by leaps and bounds, and you need to investigate all avenues before you can truly say that you can't find a plant. And then, of course, just like that silly butterfly of happiness, the minute you stop looking and are still, it will come to you. But before you succumb to sugary metaphors, try these strategies.

- Don't just scan the shelves at your local nursery and then give up when you don't see something on your list. Ask. Nurseries have a regular list of companies from which they buy, and they continually receive plant orders from growers and wholesalers during spring, and regularly at other times of the year. Maybe the plant you seek is on next week's list. If it isn't, find out if it can be special-ordered. If not, try at another nursery, which may buy from other places.

- Perhaps this is the kind of plant that won't show up at a regular, full-service nursery. Then try a specialty nursery or mail-order catalog. A Web search may turn up likely sources for your plant.
- Go to plant sales. Every year, especially in spring, dozens of organizations, such as the Washington Park Arboretum Foundation and the Hardy Plant Society of Oregon, put on plant sales as fund-raisers. Those same specialty nurseries you are trying to track down may very well be in attendance there. It's like having the results of your Google search in the flesh. And most plant sales also have a donor table, where generous members share treasures out of their own gardens.
- Visit some fabulous gardens that are open to the public. Often those gardens sell plants that have been propagated from seeds or cuttings of plants in the garden. Great plants, great deals.
- At the very least, make many other people aware of your search.
- If you haven't found the plant after all your efforts, it may be time to take the butterfly of happiness approach.

When to Buy Perennials

In a perfect world, we'd plan the garden and then buy the plants. But we know that often doesn't happen, and even if you have planned and planted, there's all that editing to do as the years go by. Even in an imperfect world, these tips on timing your purchase will help you end up with an almost-perfect garden—that is, one that makes you happy:

- When you are buying at the nursery, consider the time of year and read the tag carefully for the growing information you need. Nurseries tend to carry perennials that are in or almost in bloom. That means that you won't find many asters on the nursery shelf in April, but by August the place is sure to be awash in them.

Japanese painted fern (Athyrium niponicum var. pictum)

- Buying herbaceous plants in winter is an exercise in faith—sometimes you get nothing but a pot and a promise. In winter, you may find perennials for sale in bags of damp wood shavings. The bag contains the roots and crown (the part just at or slightly above or below the soil line). When kept cool, in the dark, and moist (but with a little air circulation), the roots will last for several weeks like this. But you don't know how long they have already been in the bag, so your first task is to pot them up or plant them out.

- It used to be that in spring and early summer, everything on the nursery table looked good, but now it seems that plants arrive from the growers just as fresh in September as they do in March. We fall for their beauty in spring, but we look for good deals beginning in midsummer, when nurseries often start putting perennials on sale.

- Mail-order catalogs offer a year-round selection of plants (unless they are sold out). Buying from a catalog also means that you have more time to peruse. Catalog descriptions are as good as a plant tag—sometimes even better, because they often include personal comments from the writer. You don't always get photos, but then, that's what this book is for.

What to Look for When You Buy Plants

Here are some tips to consider when buying perennials for your garden:

- You would think that perennials in 1-gallon containers are better than plants in smaller pots, because the gallon plants are big enough already to make a splash in your garden—no waiting necessary. But consider this: You can buy more plants if you choose those in 4-inch pots because they cost less, and sometimes plants in 1-gallon pots are just 4-inchers that have outgrown their containers and have been potted up. In other words, you may be paying for more soil, but not necessarily more plant.

- Smaller plants are easier to establish in the garden. They take less water and recover quicker if they are potbound and you must rip up the root ball.

- Buy a healthy plant. It's possible to rescue plants that are potbound and wilted, but be aware that it will take extra effort. The healthiest plants will have mostly undamaged foliage (it can be a rough life on the nursery shelf, and a tattered leaf or two shouldn't cause concern). It won't have lots of brown roots coming out of the bottom of the pot. Check for lurkers in the soil surface: little weeds, moss, and liverworts looking for a free ride home with you.

PLANTING PERENNIALS

A common complaint of gardeners is that their new plant didn't bloom when they planted it, or perhaps the floral show was less impressive than they expected. But to the plant, flowers have more than a decorative purpose; they are a means (pollination) to an end (reproduction). And flowers expend a lot of the plant's energy— energy it may not have when it is first planted. The old adage is annoying and cutesy but true: First they sleep, then they creep, then they leap. In other words, don't expect much of a show from a first-year perennial.

Planting Perennials Under Trees

Planting a perennial garden under existing trees is not impossible, but it certainly is not the easiest thing to do. Here are a few suggestions:

1. Buy small plants so that you won't have to dig a big hole in soil that will be hard and full of roots.
2. Established trees don't need irrigation except during times of unusual weather patterns (a dry spring followed by our typically dry Northwest summer), so choose perennials that prefer dry shade and you won't have to continue watering past the establishment phase—through the plant's first summer.
3. Mulch well, but don't pile it up around the trunk of the tree or the flare (the base of the trunk where it spreads out at soil line); keep the mulch several inches away.

purple coneflower (Echinacea purpurea)

When Do I Plant It?

The best time to plant perennials is when you have the time to do it. That said, we usually think of spring as planting time. The earth is awakening, our body clocks are kicking into high gear, the sun is shining longer, and plants are growing.

Spring Planting

In spring, a plant's growth is upward and outward. A healthy containerized plant will be ready to take on the new year's growth, and its root system, although not well-established, will be up for the challenge of growing new leaves. The new leaves in turn make food through photosynthesis; the food is sent down to the roots. The first year's floral show may not be as extravagant as those in later years, but if the plant is healthy, we don't mind. It is important to remember to water spring-planted perennials during their first summer; their root system isn't extensive enough to compensate for a lack of regular moisture.

Summer Planting

A hot day in July may not be the best time to plant, but sometimes it's all we've got. Planting perennials during a typically dry Pacific Northwest summer means taking a little more care about how you do it:

1. Water the container plant the day before you plant it.
2. Avoid planting in powdery-dry soil; water that, too, the day before, if necessary.
3. After planting, be sure to water the plant well (see How Do I Plant It, page 18).
4. Mulch the plant with well-composted material. Don't mulch dry soil, because it will take longer to get moisture through to the soil. Mulch is the great moderator; it slows the flow of water into the ground. That means that in winter, mulched soil doesn't get as soggy as fast as unmulched ground, but it also means that bone-dry soil that gets mulched won't take up as much water as those newly planted roots will need.

Autumn Planting

Fall is a fabulous planting time: The air gets cooler, which is less stressful on plants, yet the ground is still a warm, cozy, inviting environment for roots. Fall is also an underappreciated planting time, because there are fewer plants in bloom to dazzle us at the nurseries. Plant in fall, and you will have to water less because our rainy season begins in October.

Where Do I Put It?

The terms "sun," "shade," and "part shade" are used to guide you in placing perennials in the garden. "Sun" means the plant needs six hours or more of sunshine a day from spring into fall. "Shade" means the plant needs six hours a day of shade from trees and shrubs. "Part shade" means the plant needs a half day of sun—how's that for confusing? There are extremes, exceptions, and extenuating circumstances, of course. Usually, plants that prefer part shade grow best in morning sun, with shade from the afternoon's hotter rays. There's a difference in shade from trees (dappled) and shade from a building (complete). Full sun in the middle of a lawn is cooler than full sun up against a south-facing wall. Fortunately, there are plants that take the extremes, and this is noted in that plant's listing in Chapter Four.

It's tempting to think we can get away with planting a perennial in a spot that does not meet its cultural needs. We're tempted even further if we have ever gotten away with the practice. "But I have an Aster 'Mönch' that does just fine in shade."

Planting Perennials in Pots

There are a few differences between planting perennials in containers and planting them in the ground:

1. If you plant a container with several plants, be sure it's wide enough.
2. The depth of the container is up to you: Perennials can survive in as little as 8 inches of soil, but you'll be standing over them with the hose in the summer, watering every day. On the other hand, if you choose a pot that stands 24 inches high but want to put small plants in, you'll be wasting a lot of potting soil if you fill the entire pot and the plant only

"I never water my *Lobelia cardinalis*, and it looks great." It's true that some plants have a higher tolerance for inappropriate conditions than we may realize. It may also be that if you meet one requirement that is more important than another, you

wants 8 inches of soil. You can use a piece of fine hardware cloth (rugged galvanized screening) cut to fit snugly into the pot as the "bottom." This will provide that all-important drainage yet keep the soil in the top 10 or 12 inches of the pot, so you won't be washing away the plant's environment.

3. When you plant, set the plants so that you fill the pot up to within an inch from the rim. That way, when you water the container, the water won't all run out over the sides.

4. Pots dry out faster than the ground, so your main concern with containers will be summer watering. Terra-cotta dries out quickly in warm weather; check these pots daily. Glazed or water-sealed terra-cotta retains water longer. Faux terra-cotta pots are looking more stylish all the time and they are a cinch to use: They dry out more slowly and are lightweight.

5. In winter, your main concern will be keeping the pots from being waterlogged. Pot feet help keep drainage working. Drawing pots up under the eaves of the house helps, too. If there is the potential for below-freezing temperatures—especially days filled with freezing, dry, desiccating winds—then wrap the pots with layers of newspaper or bubble wrap or take them into an unheated garage until the weather has settled down.

6. The soil in the pot will decrease as time goes on, so perennials that you keep permanently in the container will benefit from having fresh soil added. Or take the plants out and refresh the entire container.

can keep a plant happy and growing. For example, some shade plants, such as astilbe, will tolerate a fair amount of sun, if provided with regular (some might say, copious) amounts of water.

Toying with a cultural requirement isn't exactly playing with fire, but then, you won't get a lot of sympathy when you whine that your *Brunnera* 'Jack Frost' withered when you planted it in your full-sun rockery. Perhaps it would be better to keep 'Jack Frost' in a pot until you've found a nice shady spot. Maybe you could plant up a nice container that would work well in a shady entryway. Combine the herbaceous *Brunnera* with an evergreen such as one of the hellebores (*Helleborus foetidus* has fingerlike leaflets that would look good in contrast to the Brunnera's wide, solid foliage). Stick a few hardy cyclamen in for a shot of pink.

How Do I Plant It?

Generally, the planting instruction for perennials is simple: Just stick it in the ground.

But then we look at how the plant grows in the pot and wonder if it's OK to cover up any stems with soil. Then we take it out of the pot and see that its root system is one solid chunk of brown roots circling the inside of the pot, and we wonder: Should I do something about that? What about fertilizer and mulch in the planting hole? See, there's more to it than you thought.

1. Because you should plant in moist ground—not wet and not powdery dry—you may need to water the ground the night before if the weather has been dry.

2. When you plant an entirely new garden, you can dig one giant hole for all the plants and prepare the ground beforehand. When you are adding a plant or two here and there in the garden, dig smaller holes for specific plant sizes. If you are planting a perennial from a 4-inch container, use a sturdy trowel to dig a hole; if you are planting from a 1-gallon pot, the best tool for digging among existing plants is called a poachers or planters spade. It has a long, narrow scoop or blade (the part you stick in the ground) and is less likely to damage neighboring plants. Use one with a short, D-shaped handle, and you won't bump the handle end into a nearby fence or tree. Dig a hole that's wider and a little deeper than the pot. This loosens up the soil and gives the expanding roots an easier time of it. Mound up the soil in the hole a little bit, so that you don't bury the crown of the plant.

3. It is important to examine the condition of the plant's roots when it comes out of the pot. A potbound plant will have little soil left in the container; instead, the roots will have filled up the space. The brown roots you see circling around are not doing the plant much good. The white, healthy roots grow tiny root hairs that

pull in water and minerals for the plant. Use a hand cultivar to comb out the roots, loosening the solid mass. You can trim off excess brown roots (several inches' worth is OK) with scissors or hand-pruners. The roots of a plant in a small pot can be loosened by hand.

4. Place the plant in the hole so that its crown ends up just an inch or two above soil level.

5. Fill the hole around the plant with soil and water it well, adding more soil where you see pockets of soil sink. With trees and shrubs, it's important to backfill with the same soil that you dug out of the hole, but with perennials the rule is relaxed somewhat. You can backfill with soil mixed with compost if you want, but you're making more work for yourself. Instead, see the next step.

6. Mulch after planting. Apply a layer of well-composted material such as composted yard waste or washed dairy manure. "But wait," you say, "I have heavy clay soil and I wanted to make it better by adding compost to the hole." Fine— mix and backfill, mix and backfill. But it's easier to wait until you have finished planting and add the mulch; it does the same job. Or, rather, the little fauna of the soil do the job for you. As the compost breaks down, it is taken into the soil and the soil texture is improved. Let those worms do their work!

7. Gardeners can't keep their hands off the fertilizer bag. In general, the garden can get all the nutrients it needs from the soil and—here it comes again—that layer of well-composted organic material, the mulch. However, some needy plants—roses, for example—do need to be fertilized after planting, and lilies and delphiniums perform better when given extra nutrients.

globe thistle (Echinops ritro)

Chapter 2:

DESIGNING WITH PERENNIALS

What are the fundamentals of design when it comes to using perennials in the garden? There are tangible components of garden design, such as hardscape and water features, and there are intangible features, changeable components such as the quality of the light and the shape of a plant because of weather conditions or growing aspects.

The first consideration in your garden design should be you. How do you move about the garden? This determines pathways and patios—the so-called hardscape of the garden. Once you know where the plants won't be, you can think about how to use them where they will be.

Gardens—other than their hard surfaces of stone and brick—are alive, and therefore we can't design them as we can a living room or bathroom. Not that we don't try, of course. Books and magazines are full of garden plans that you can copy. But even in the most well-laid garden plans, plants happen.

HOW TO CHOOSE THE RIGHT PLANT FOR EACH LOCATION

The overriding factor in designing with perennials is where to put the plants so that they will be healthy, because healthy plants are good-looking plants. Not only is a healthy garden a beautiful garden, but when you make good choices of plants and placement, your garden will result in less work and fewer problems. And if you have to spend less money on water and controls for pests and diseases, you will have more money for buying plants. Works out well, don't you think?

Get to know the specific characteristics of the different areas in your garden so you can create a design that pleases you. This is not difficult. Rather than focusing on which colors to combine, begin with reading the cultural notes for each plant!

For instance, you love hardy geraniums but think that they grow only in full sun. However, the mourning widow (*Geranium phaeum*) appreciates part shade, and the evergreen *Geranium macrorrhizum* takes even dry shade. Or say you love agaves, but your garden is mostly shady. Then you'd better love agaves in pots, so that you can set them on the driveway against the south-facing garage door.

Avoid combining plants with contrasting cultural needs, because this leads to stressed plants that are susceptible to problems—and then there goes your perfect design. For example, if you can't do without a water-loving astilbe in a shady location, combine it with *Rodgersia*—another shade plant that needs regular water—rather than *Epimedium* or hardy cyclamen.

HOW TO DESIGN A GARDEN

You don't have to fill your head with rules when you decide where to put plants—you probably use design concepts without thinking. So consider the following sections simply as guidelines. If you have the space and if yours is a new garden, you can move nursery pots around the area, stand back, and squint (which helps you imagine the plants all grown up) to see if you like your choices.

giant scabiosa
(Cephalaria gigantea)

Using the Elements of Design

The elements of design—whether for gardens or any other composition—include these components:

- repetition
- variety
- balance
- emphasis
- sequence
- scale

Running as a common thread through each of these design elements is a subset of characteristics that include:

- line
- form
- texture
- scent
- color

The design elements and their characteristics are not separate unto themselves; rather, their importance and their use overlap and blend together. Here are some ways to use these design elements and their characteristics with perennials in the garden.

Repetition and variety would seem to be design elements that are at odds with each other, but actually each can be used to balance the other's impact. Repetition of specific plants helps unify the garden and keeps it from looking like the multicolored sugar sprinkles on top of a cupcake. But far on the other hand, repeated ranks of the same plants can begin to look like a public landscape—and you don't want your garden compared to parking islands at the mall.

The middle ground is to repeat the same plant occasionally but also to pick up on its color, form, or texture and repeat that characteristic throughout the garden for variations on its repeated theme. For instance, to repeat the low, hummocky form of Geranium macrorrhizum, use the rounded form of lady's mantle (*Alchemilla mollis*). Or repeat the pink flowers of *Linaria purpurea* 'Canon Went' (toadflax) in early summer with the pink show of *Aster* 'Andenken an Alma Potschke' in late summer—and both are tall and thin, which repeats not only the color but the form, too.

The preceding example shows how the elements of design overlap: In addition to being

Whether your garden came with existing hardscape—fences, walls, patios, and paths—or you are designing your own placement, you can use perennials to accentuate, soften, and integrate the structures. Take, for example, the intersection of two concrete paths. For a formal look, plant a small catmint (*Nepeta* 'Little Titch') in each of the four 90-degree angles of soil. If these seem too blousy, fill in the corners with small, compact mounds of thrift (*Armeria*). At the base of an arbor, plant a *Geranium renardii* to hide any necessary fixtures for the structure. Call attention to a piece of stained glass set into a gate by planting a perennial that blooms in the same color. Perennials are much easier to move than a concrete patio, so use plants to help maintain or change a look, before resorting to the sledgehammer.

thousand-flowered aster (Boltonia asteroides 'Snowbank')

a means of repetition, color and form also have much to do with emphasis in the garden. A contrasting spot of yellow draws the eye. Accent a particular area with the tall, upright form of *Verbena bonariensis*.

The succession of flowers described in the toadflax-and-aster example also shows how sequence

The element of scale involves the impact of your house and other buildings on your garden design. Large, lumbering shrubs overwhelm a tiny house and garden. A landscape with only dwarf, rock-garden plants looks like it was designed for a Barbie doll. Use a variety of plant sizes to ensure that the plants neither over- nor underwhelm the site.

These design elements must be fit into your own space. No one can tell you what to do (unless you hire someone to design your garden for you). Magazine and book designs, with their neat little blobs of color here and there, all correctly labeled, don't translate well into every single garden. By all means, try such designs on for size, but be ready to substitute plants and placements when you discover that there are no cookie-cutter gardens.

Using Colorful Foliage

Foliage—striped, blushed, edged, mottled, or downright saturated—is big in perennial gardens these days. To some, it's the height of good taste; to others, a green leaf stippled in yellow looks diseased. For those gardeners who fancy it, variegated and colored foliage add not only interest to

can be integrated into your garden design, which can change the impact of emphasis as well. The ebb and flow of different parts of the garden through changes in foliage and flowering over time yield emphasis now here and then there.

a garden or container but also can carry on a plant combination far longer than flowers.

However, as with every fashionable thing in this world, too much of a good thing loses its impact. Instead of going overboard on the variegation, be conservative. Judicious use of variegation makes a plant stand out against a quieter background. It's much more effective to use one variegated plant in a sea of green and flowering plants than to have the entire scene taken up with it.

For example, the cream-edged leaves of the variegated wallflower *Erysimum linifolium* 'Variegatum' blend into the scene if planted alongside *Carex morrowii* 'Variegata'. You'll end up with zebras on the brain. But take the wallflower and plant it with *Bergenia* 'Appleblossom' and some black mondo grass, and the variegation pops out, providing a powerful punch.

Choosing a Garden Style

The elements of design can be used in any garden style, and perennials—with their huge variety of forms, textures, flowers, and growing habits—fit into any garden style. Below are some of the more common styles.

lamb's ears (Stachys byzantina)

- Cottage garden, an informal style full of perennial and annual flowers.
- Formal garden, where the outline is distinct and straight and where hard lines are softened by plants.
- Naturalistic garden, in which sweeps of plants are important, as are ornamental grasses, color tones come not just from flowers but also from foliage as it changes season to season. The naturalistic style is a movement that has come to the United States from Europe.

Which is your style? You may admire the strong lines of a formal design but prefer to relax amid an informal stand of foxgloves and hollyhocks. You may find bunches of ornamental grasses a tad too wild and long for straight lines with geraniums billowing into the paths. Visit many gardens and figure out what you like. Once you have determined which garden style you prefer, choose and combine plants to suit.

HOW TO DESIGN A MIXED GARDEN USING NONPERENNIAL COMPANIONS

The mixed garden is a joy. Trees and shrubs provide those all-important "bones" to the garden—structure that carries through year-round. In winter, the mixed garden is not a wasteland of dead stems and seedheads (as important as those are for birds); instead, trunks and branches take center stage, with some winter blooms as accents for emphasis.

For example, a carpet of hardy cyclamen (*Cyclamen coum* blooms in winter) under the shaggy stems of an oakleaf hydrangea is a hearty sight in cold weather and stresses both variety of form and sequence of flowering.

Perennials need these partners in the garden in summer, too. It helps them to show off—employing the design element of emphasis. As a backdrop, the leafy greenness of nonperennial companions sets off colorful perennial flowers and foliage; the companions provide shade as well as a frame.

Hedges—short or tall—offer a good green curtain for perennials. For instance, Japanese

Designing with Perennials in Pots

Perennials can live for years in containers if given the proper care, so feel free to decorate your patio, balcony, or deck with them. You can combine perennials with woody plants in large pots, but remember that trees and shrubs may eventually need to be planted out in the garden. Here, just as in the garden, it's important that you group plants according to their needs; no sense in overwatering one plant and underwatering another when they are in the same pot.

There are no approved and disapproved lists of perennials for pots. As long as you care for them properly, all will live long and prosper. Some plants you may want to particularly choose for pots—the aggressive ones. If you don't like the thought of Mrs. Robb's spurge (*Euphorbia amygdaloides* var. *robbiae*) expanding its territory in the ground yet you like its evergreen appearance, keep it in a pot.

Pots are perfect for decks, patios and front entrances. They soften hardscape, but also dress up the scene, especially when you choose attractive containers. Large containers can stand alone, but small containers look best when grouped for effect. Cluster pots of perennials in opposite corners of the deck or patio. This gives you the opportunity to spotlight perennials in bloom, and reorganize the pots for another, later show. Use one large pot and two or three baby pots at its base for a front-door arrangement—if it's in the sun, you could choose tall and short bearded iris. Plant a formal concrete urn with short varieties of sedum or billowing soapwort. Place single, tall pots within the garden to give height to a bed, and add something spiky such as *Crocosmia* or *Schizostylis* for drama.

primroses (Primula)

spurge (*Euphorbia griffithii 'Fireglow'*)

holly (*Ilex crenata*) includes cultivars with a year-round clean look, such as 'Helleri' (to only 3 feet) and 'Jersey Pinnacle' (to 6 feet). Boxwood, another example, is the classic choice; use a row of the low English boxwood (*Buxus sempervirens* 'Suffruticosa') to line a bed or walkway, setting off the flowers and foliage within. Or choose a flowering evergreen shrub such as Escallonia for its greenery and for its small clusters of rose-pink flowers late into the year; this employs the element of sequence.

Shrubs and small trees play a part within the mixed garden, too, for emphasis and protection from weather. For instance, the deciduous shrub *Enkianthus campanulatus* is a piece of architecture: It grows stiffly upright but its branches grow on a horizontal plane, providing contrast to mounding or short, spiky forms below it—variety of form. Its small leaves don't throw too much shade, and so beneath its branches you can place perennials that need sun to part shade and that tolerate slightly acid soils—lungworts, for example.

Light and airy shrubs fit well into the mixed garden. Consider, in a sunny spot, a combination of the small mock orange *Philadelphus* 'Belle Etoile', ground cover roses with Oriental poppies, and viburnums with asters. For part shade, try the royal azalea (*Rhododendron schlippenbachii*) and the buttercup hazel (*Corylopsis pauciflora*) with lungworts (*Pulmonaria*).

HOW TO ADJUST THE DESIGN OF A MATURING GARDEN

Maintaining a particular design by keeping tight control of a landscape may be possible in a public garden or a historic garden, but home gardeners don't have the time or the staff for such an effort. And when another homeowner moves in and gets used to the garden, then it's time for a change. Garden designers of the eighteenth century created composed landscapes made to look like paintings, but even their composed settings aged and changed. Ultimately, we should be ready to reconnoiter, reorganize, redesign, and replant.

Plants grow up; for instance, phlox can expand its holdings until it takes up twice the space it started with. Plants that increase in girth can be divided (see Chapter Three), but a stand of Siberian iris makes a formidable impression, so don't be too hasty to chop plants up into smaller bits of plants just because you think you should. Mature gardens, with plants amiably jostling elbow to elbow, give a pleasing, bounteous look; perennials growing with lots of space between them can look like a scientific experiment.

Short-lived Perennials

Perennials have variable life spans; they may edit themselves out of the garden because of conditions or because of their own genetic makeup. For example, some plants, although they continue to bloom, never set any seed—which, for a plant, is the whole point of blooming; these plants can bloom themselves to death. You can decide to replace such a plant with another identical species, find a different plant for the space, or let neighboring plants fill in.

The wallflower *Erysimum* 'Bowles Mauve' is just such a plant. It quickly makes an impressive 3-foot mound of bluish-green foliage and begins blooming almost immediately. Clusters of deep mauve flowers appear on narrow stems. The flowers keep coming, sometimes all year in mild Pacific Northwest winters. Then, after about three years, you see a stem die off. Within a short period of time, the plant either dies off completely or you wish it would. You didn't do anything wrong. It's the nature of the plant. You can replace it with another 'Bowles Mauve' or not.

Self-seeding Perennials

Plants that set seed can be a boon and a bane at the same time. On the plus side, they self-seed new plants all around them, saving you the task of planting, but on the negative side, these volunteers may grow where you don't want them.

For instance, the maroon-black flowers of *Knautia macedonica* are on long, flexible stems that lean this way and that until they weave several plants together into the scene. It's difficult to cut back these interwoven flower stems that still have flowers opening high up on the stems, so old flowers farther down the stem begin setting seed and *Knautia* reseeds from these. In two years, one plant can increase to several, forming low, wide rosettes of leaves that tend to muscle out small plants. But because it's a plant that looks better with neighbors close by, rather than being alone in a spot where it merely flops, you must choose where to let it self-seed. In late winter, it's easy to decide which Knautia plants to leave and which to remove. A little help with a hand cultivator or digging fork, and the plant pops out of the ground.

The Mixed Garden

The trees and shrubs in the mixed garden do their part in changing the design of a maturing landscape. A birch tree grows up; a mock orange gets middle-age spread. What was once a sunny terrace now has dappled shade from the expanding plum tree.

Or a much-loved flowering cherry has died and now the woodland garden is a desert. Blazing full sun doesn't suit a plant such as delicate meadow rue (*Thalictrum*), chosen for the shade of a now-missing tree, and the sudden change between shade and sun can wreak havoc on even those plants that might grow fine in full sun if they've grown accustomed to it slowly.

Ultimately, whether trees and shrubs grow larger, increasing the shade beneath them, or are removed for whatever reason, suddenly eliminating the shade they once provided, your perennials may need moving too. Perennials, by and large, are an accommodating bunch when it comes to transplanting. They'd rather be moved than put up with adverse conditions. So be prepared to change your garden design as your garden evolves.

gaura (Gaura lindheimeri 'Whirling Butterflies')

Chapter 3:

MAINTAINING THE PERENNIAL GARDEN

The problem with Pacific Northwest garden-ers taking advice from national sources is that the timing of garden tasks here, as well as our gardening styles, are different. Each region's climate dictates when we must, should, or don't have to be occupied in the garden. In colder regions, the garden is frozen in winter; here in the Northwest, we have one or two cold snaps that, depending on their severity, may kill off some plants and force others into dormancy. In warmer climates, plants get even less rest than they do here.

Although in the Northwest the only year-round garden occupation is weeding, there are always activities we can find to do, which means we can spend a particularly beautiful day outdoors in the garden even in winter. We can spread a winter mulch almost anytime during the dormant season, because there is seldom a snow cover that lasts for long. Because spring growth often begins before the official date of the spring equinox and we can see flowers late into fall, the Northwest garden always beckons.

A Glossary of Garden Maintenance Tasks

Gardening terms such as cut back, cut down, pinch back, shear, and deadhead, left undefined, cause consternation among gardeners wanting to learn just how to take care of their plants. Here is what they mean:

Cut back: Cut off part of a stem. Do this when a plant is finished blooming, and you can encourage a second flush of flowers later in the season. When you cut a stem back, you cut below the last (finished) flower and just

JUST HOW MUCH IS THERE TO DO?

How much work does a perennial garden require? The answer is: How much work do you want it to be? And, anyway, aren't you a gardener because you love gardening?

No landscape is maintenance-free, except maybe artificial turf and plastic flowers. Having

above where you see the potential for new leafy growth (often in the leaf axil, which is where the leaf joins the stem). You also cut back plants early in the season to encourage branching (such as *Penstemon*), to make them bloom when they are shorter (such as *Campanula*), or so that they don't spread out, leaving a hole in the middle of the plant (as in *Sedum* 'Autumn Joy').

Cut down: Cut the stems close to the ground. This cleanup activity is done in the dormant season or with plants that go dormant during the growing season (Oriental poppies, for example).

Pinch back: Take the stem off, but only as far down as the first or second set of leaves. This is often done on new growth, which is tender enough to do with your fingers, hence the term pinch.

Shear: Indiscriminately and uniformly cut back, often using hedge shears. If you don't want to play hunt-and-find with the spent *Coreopsis* 'Moonbeam' flowers mentioned under High-Maintenance Gardening Style (page 38), you can shear the whole plant back a few inches.

Deadhead: This is not a fan of the rock band, but an activity carried out during the growing season. Cut off just the spent flowers. This encourages some plants to continue blooming. Pincushion flowers (*Scabiosa*) can be deadheaded for a longer blooming time.

trees and shrubs means you rake leaves and prune. With perennials, you cut back to control growth, cut down spent plants, and groom plants after flowering. Those who prefer a garden as neat as a pin must cut back again, cut down several times, and groom constantly.

And what is your reward? The garden on a warm summer day. Both you and the bees hum lazily as you wander from poppy to rose to *Rudbeckia*. The bee is more industrious than you, collecting its pollen. You are more likely to end up like the resident garter snake, asleep on a rock in the sun. Make that a rock for the snake and a chair for you.

Maintenance tasks depend on the plant, so read about each plant's care in Chapter Four, Perennials from A to Z. But there are general guidelines for upkeep, which we can discuss here.

Below are some ways of looking at a garden maintenance schedule. Your style will most likely fluctuate among these three scenarios, as fits of tidiness correspond to empty spaces in your calendar.

Low-Maintenance Gardening Style

If puttering about in the garden is not a full-time occupation for you, if you have little time to garden but love flowers, choose perennials with an evergreen presence, such as Mrs. Robb's spurge, and evergreen grasses, such as the orange sedge (*Carex testacea*), and then accent with color from reseeding perennials such as toadflax (*Linaria purpurea*).

Perennials that need more than one thing done to them, such as the mourning widow, *Geranium endresii* 'Wargrave Pink' (it must be cut back in winter and cut back again after blooming), may not be for you. Perennials that need staking, dividing often, or treating regularly for any sort of pest or disease problem—no matter how lovely they are—are not for you.

Optimal-Maintenance Gardening Style

For gardeners who love to be in the garden and don't mind the work but still have a real job, the world is yours. Perennials don't need the amount of work some people think: an hour or two of tending a week, cleanup at the appropriate time of year, and let the garden do what it does best. For you, the best plant support is another plant, so you don't spend time staking and tying. And remember that the seedheads you leave on the phlox and goldenrod provide perching places in winter for finches, chickadees, and bushtits. Your garden is enjoyed by everyone.

High-Maintenance Gardening Style

You spend every possible waking moment in the garden and think nothing of spending half an hour snipping off the spent blossoms of *Coreopsis* 'Moonbeam' so that the new flowers will show to their best effect. You like to keep a tight watch on plants. No flopping is allowed in your garden, so plant supports go in early for peonies and delphinium. Daisies and sedum get cut back by a third to a half in May so that they will stand at attention when in bloom later in the summer. There are no "messy" plants for you, because you are on top of cleanup. Brown stems might as well be outlawed.

SPRING TASKS

The rains taper off, the sun peeks out again, and we are drawn, inexorably, to the out-of-doors. There's so much that we want to do in the garden that we find ourselves constantly distracted. We head outside to weed one corner of the garden but see the emerging stems of astilbe and stop to take a look. We find that a little cleanup is in order here, so that we'll be able to admire the frothy pink flowers later on. Did we forget to clean up the old hellebore leaves? Wouldn't this be a great place for that new *Epimedium* that's waiting on the front porch? Soon our original goal is forgotten, and we happily lose ourselves in whatever is at hand.

In spring, the gardener wants to get things in order. Spring planting proceeds apace, especially because there are plant sales every weekend offering choice selections from specialty

wallflower (Erysimum 'Julian Orchard')

nurseries, and full-service nurseries are stocking shelves with delectables.

Your garden's daffodil show was beautiful in late February and early March, but now you're getting just a little tired of all the daffodil foliage. You know you aren't supposed to cut it back—the leaves are photosynthesizing and sending food down to the bulb, creating energy for next year's

Fertilizer versus Mulch

We automatically reach for the fertilizer in spring. Don't our perennials need a boost? What would they do without our annual application of fertilizer?

They would probably do just fine, and they might do even better if we lay off the fertilizer and rely on organic matter to do the job. Organic mulch provides enough nutrients for plants to grow well and flower, and it's in only extreme cases that you may want to add to this by sprinkling on some fertilizer.

Perennials in pots do not have access to real soil and so may do better with a light dusting of some all-purpose organic fertilizer in spring. Some gardeners prefer to help along plants that need more acid or more alkaline soil, and so an amendment helps here, too. You may want to apply some dolomite lime to the hellebore planting hole (even just a piece of chalk) to increase the alkalinity, or sprinkle leftover coffee grounds around the lungworts after you've cut back the spent flower stems. But in general, the only fertilizer you need apply is a good mulch.

show. Most experts advise against braiding the foliage, as this can damage the leaves, which then won't be able to photosynthesize. Instead, carefully fold the leaves down and hide them under some big-leaved perennials and ground covers. Bergenias, coral bells, mounds of geraniums, and lady's mantle can all help.

The big, wide, floppy foliage of hybrid tulips is harder to deal with. But showy midspring bulbs, including the giant Darwin hybrids, work best in pots anyway, because they don't like any mollycoddling during the summer. Tulips in containers allow you to move them backstage at the end of their show, where they can spend the summer as dry as they like it.

Nipping Pest Problems in the Bud

There's more to do in spring than plant. It's time to be on the lookout for spring pests. Using mechanical controls and encouraging wildlife in your garden—a few of the garden helpers—keeps your perennials healthy at a low cost and with less work from you.

Cutworms

One of the worst spring pests is cutworms. These can eat up seedlings and many kinds of perennials that have a lot of basal foliage, such as foxgloves, before you know it.

Keep watch. Do you see leaves getting munched on? Dig around lightly in the soil or cut off the lowest leaves. Cutworms are grayish brown in color; they curl up in a "C" when disturbed. They usually feed at night.

The best cutworm control is immediately dispatching them with your trowel, weeder, pruners, or garden scissors. Go night hunting with a flashlight. Encourage predators of cutworms, which include birds and spiders. It might help to put the bird feeders close to plants that might be eaten by cutworms. Protect seedlings with paper collars that are pushed slightly into the soil; keep these on until the plants are big enough to fend for themselves. Larger plants may be disfigured but won't be killed. Cutworms are a problem in spring but turn into moths later in spring.

Aphids

And then there are aphids. They come in green, brown, or black and can cover the stems of

baby's breath (*Gypsophila paniculata* 'Pink Fairy')

pincushion flower (Scabiosa caucasia 'Fama')

succulent new growth on perennials as well as annuals and shrubs. In our mild maritime Pacific Northwest climate, aphids start early in the year and continue until cold weather. Some plants get aphids even in the winter. Aphids debilitate plants but don't usually kill them outright.

Spraying insecticidal soap is a low-toxic way of controlling aphids, but it's even easier to knock the aphids off with a sharp spray of water. Once knocked to the ground, they don't return (it's another generation that shows up next time). Keep aphid activity down by reducing or eliminating the amount of fertilizer you use. Overlush plants are an all-you-can-eat buffet to aphids.

Again, wildlife is your friend here. Birds, especially the little bushtits, can pick a stem clean of aphids. Soldier beetles, which are shaped like capsules and have black wing coverings and orange heads, feed on aphids. Ladybugs eat aphids, but ladybug larvae devour them. Ladybug larvae look like tiny black or black-and-orange alligators. Look up a photo of ladybug eggs, so that you won't inadvertently destroy them. The yellow eggs are laid in clusters of ten to fifty, and look like little footballs standing on end; they are usually on the undersides of leaves.

Slugs

The silvery sheen of a slimy trail leading to and from a ravaged *Hosta* can only mean one thing—the gardener's nemesis has been at work. The brown European slug destroys succulent new growth of some of our favorite plants. (Northwest native slugs eat dead plant material and are not considered pests.) Chemical controls are not the way to go, as they are toxic to wildlife, pets, and children.

The most satisfying way of getting rid of slugs, of course, is to do the deed yourself, but you would have to be on slug patrol for too many months of the year. You could pay the kids a nickel a slug (make the kids wear gloves!). Beer traps—cheap beer in cat-food cans that are set on the ground—are effective, but rather unpleasant to empty. There are safe controls available on the market today, using pellets of iron phosphate. If you haven't used one of these products before, you'll be amazed at how well they work. No more holes in the hostas.

Controlling Floppy Plants

The different growing habits of perennials make for variety and visual interest in the garden. Some plants grow up, others out; some creep along the

Saving Seeds

Want more granny's bonnets? Wish your Welsh poppies were by both the front and back doors? You can save seeds from many of your garden perennials to share with friends or spread around your own garden. Seeds need to be fully ripe to be viable (able to germinate), and that means leaving the spent flowers alone until the seedhead looks dried out. You know when the seeds of the poppy are ripe, because you can shake the poppy head and hear them. Of course, you're also shaking the seeds out through tiny holes, so do this over a bowl or into an envelope.

Save the seeds in a dry envelope or film canister in a cool, dark place. Or plant them out immediately in pots or in the garden. You may get some genetic variation in plants—such as varying flower colors—but that just adds to the interest. Here are some perennials that are easy to grow from seed:

1. rose campion (*Lychnis coronaria*)
2. jupiter's beard (*Centranthus ruber*)
3. lady's mantle (*Alchemilla mollis*)
4. granny's bonnet (*Aquilegia vulgaris*)
5. tickseed (*Coreopsis grandiflora*)

ground. And others flop. This is of great concern to many gardeners, who don't want their *Sedum* 'Autumn Joy' to spread open in the middle. To stake or not to stake? Or is it possible to limit floppiness with pruners instead of stakes and string?

Many plants can be cut back by one-third to one-half in spring to keep them shorter yet still allow them to bloom. Sedum, *Eupatorium*, and coneflowers (*Echinacea*) will all take cutting back and still bloom later. Cutting back can also be done just for controlling the bloom time and not the plant's size. Garden chrysanthemum can be nipped back twice to make sure they bloom in late summer and not earlier. Globe thistle (*Echinops*) can be cut back hard after it blooms once and will flush out with new growth and a few more blooms. Cut off the old flower stalk of a delphinium and you'll get more flowers.

But when it comes to floppiness, many gardeners choose to plant perennials close together and then allow them to have at it. This creates some surprising and lovely combinations at various times of the year. An orange tiger lily leans over into the catmint (*Nepeta*); a stem of deep maroon purple Knautia flowers wanders among the *Crocosmia* 'Lucifer'. New pictures are created daily.

SUMMER TASKS

You need a rest. Let the summer garden take care of itself. But if you feel compelled to work, you can always deadhead. Cut the oriental poppies down when you've let the seed pods ripen. Spot-water the plants that need it in only the driest part of the summer—the mulch you applied in winter will be a big help here, reducing the need to water.

Visit other people's gardens. Make mental notes about which plants to move and which new plants to buy when fall comes. Do this while sitting in a lawn chair with a cool drink in hand. Take a nap.

AUTUMN TASKS

Yes, fall is a fabulous time to plant. The nursery pots that have been accumulating on your deck or porch all summer long, as you visit this nursery or that plant sale, can now go in the ground. The warm soil and autumn rains will encourage good root growth well into winter and get your plants off to a robust start in spring.

Cleaning Up

We hear way too much about fall cleanup. October and November may be a good time to clean up the garden in Minnesota, but there is still too much going on in our Pacific Northwest gardens. We must wait until late fall—that is, the first half of December—to cut back those plants that have died back.

Some disease-prone plants do best if their foliage and stems are removed as early as possible. This can keep peony botrytis at a minimum. But because the foliage of many peonies takes on lovely apricot tones in fall, you may want to delay the immaculate cleaning if you don't have a botrytis problem.

soapwort (*Saponaria ocymoides*)

Remember that leaving the stems and seed-heads of plants provides good habitat for birds. Finches perch on goldenrod. Bushtits hang upside down on a Japanese anemone. Chickadees flit through the penstemon. These birds are a great asset to our gardens, and it doesn't take much to give them a place to hang out and eat.

Moving Plants

The plants that you let go to seed over the summer may now have a flock of seedlings sprouting at their base. You can pot up seedlings of hellebores, columbine, lady's mantle, and Welsh poppies now or during mild winter weather (see Dividing Perennials, below, for details). If you don't plan on replanting the seedlings immediately, keep the pots in a sheltered place for the winter, or give them away to grateful gardening friends.

Dividing Perennials: Making More of a Good Thing

The dormant season—late fall and winter—is also a good time to divide plants that have become congested (which may lead to a lack of flowers) or

Overwintering Perennials in Pots

It's only the perennials in pots that you should have any concern for in winter, because those plants are susceptible to becoming waterlogged and/or freezing their most vulnerable parts when they are exposed to cold.

What kills containerized plants during winters in the maritime Pacific Northwest is wetness. Pots that sit flat on the ground—whether that's a concrete patio, a wood deck, or the soil—are likely to accumulate too much water, even with the required drainage hole in the bottom. They can't drain the rain quickly enough during our rainy winters. When the air spaces in soil that are vital to a root system's survival are filled with water instead of air, the roots suffocate.

A dish under a pot is a good idea in summer. The water that runs out of the drainage hole sticks around to be reabsorbed. The air is dry, and the water evaporates instead of keeping the soil saturated. A dish under a pot in winter, however, turns the pot into a swimming pool.

So preparation for winter in the potted perennial garden includes taking dishes out from under pots and making sure that there is adequate space for drainage. Many gardeners buy cute pot feet and put three under each pot, successfully providing drainage by adding a couple of inches of space between the bottom of the pot and the ground. Special purchases do not have to be made for this winter accommodation, though; you can always set pots up on two boards or a series of bricks.

Or bring the pots in out of the rain. Moving the pots under the eaves may seem a daunting task if your container garden includes lots and lots of pots, but for a few containers (especially lightweight, faux terra-cotta), it's within reason.

Another potential threat to plants in pots in winter is the cold. Because roots normally grow underground where they are well protected from harsh elements, we don't think of them as being vulnerable. But in fact it is the roots of a plant that are the most tender, and when there is only an inch of terra-cotta between them and a bout of 10-degree days, there could be trouble.

There's strength in numbers, so the first thing you can do to prevent damage to plants in pots is to group the containers together—preferably up against the house, under a patio awning, or under the deck.

are crowding out other plants. Look at the way the plant grows to figure out how to divide it.

Plants such as hostas can be cut straight through. Hostas are good candidates for dividing, although you may need to take a saw to cut the pieces apart.

Plants that increase by growing new rosettes of foliage next to the mother plant, such as thrift (*Armeria*), can be separated. It might be easier to dig up the whole plant to separate them; pull apart and cut connecting roots.

Plants that expand by stolons—underground stems—such as *Phlox paniculata* can be dug up in sections, slicing through the root system as you go.

WINTER TASKS

Winter is pretty much vacation time for the gardener, especially when it comes to perennials. Those that keep any sign of life aboveground (such as a rosette) can take care of themselves.

Autumn cleanup can now be considered winter cleanup. Cut back dead foliage and stems. A good rationale for waiting until late winter to do this is because the dead leaves and stems offer some protection to the crown of the plant from excessive water and cold.

Winter is a good time to plan for summer dryness. Snake a soaker hose through the areas of the garden where you have grouped those plants that need supplemental water during dry weeks. Or rearrange the soaker hose that you have, if you have shifted the perennials in your garden around. If you don't do it now, you won't get it done.

During mild weather, dig, divide, and pot up extra plants to give away or donate to a spring plant sale.

Wait for the first flower. You won't have to wait long, as the hellebores will already be peeking out of the ground.

A Good Time to Mulch

The easy answer to the question, "When is a good time to mulch?" is: "Whenever you have the time." Many of our garden tasks get done when we have a chance to do them, regardless of advice, calendars, and what the neighbors are doing. That's life.

Fortunately, a good time to put mulch down is in the quiet garden season—winter. In winter, the mulch can be piled up a few inches on top of dormant perennials with no harm done. In winter, the soil is moist, so you're already planning for the dry days ahead by laying a protective blanket. Throughout the rest of the rainy season, the top layer of soil doesn't get beaten down so that it forms a crust when dried out. Drenching rains don't waterlog soil beneath, because the mulch slows the flow of water by trapping drops within its maze of various-sized particles. The seeds of winter weeds get covered up and so don't have light to germinate.

But what if somehow winter got away from you without the mulch being spread? The world is not going to come to an end. Spring is a fine time to mulch the garden, although you do need to be a little more careful about those emerging plants.

Actually, almost anytime is a good time to mulch. The only part of the year you want to avoid is full summer, when the soil is bone dry. You don't want to protect dry soil, and you don't want to be out there on a hot August day shoveling manure, do you?

Nectaroscordum siculum

Chapter 4:

PERENNIALS FROM A TO Z

→ More Than 150 Genera Described

→ Species and Cultivars

→ Garden Use and Cultivation

For your gardening pleasure, here are more than 500 different perennials described for Northwest gardens. Within each genus listing, you'll find the botanical name, common name (if the plant has one), and the plant family. Following that is a general description of the genus, and sections on cultivation (how to grow) and garden use (where and with what to grow). Next, you'll see descriptions of specific plants and cultivars, including time of flowering, size, and hardiness. (Note: Dimensions are given in height x width; a single measurement refers to height unless noted otherwise.)

Most perennials will grow well in a variety of soils (neutral to slightly alkaline or slightly acid) as long as the soil is well drained. Of course, there are persnickety perennials—those that grow well only if particular requirements are met. Some species of *Lysimachia*, for example, prefer damp to wet soils, while the mint hyssop *Agastache rupestris* needs sharp drainage. Pay close attention to these particulars, but know that there is a wide selection of plants with accommodating natures.

Acanthus spinosus

Acanthus

(bear's breeches) *Acanthaceae*

Bold plants with large, often glossy leaves. Spikes of subtly colored, foxglovelike flowers rise above leaves.

CULTIVATION

Sun or part shade. Provide good air circulation to keep powdery mildew at bay.

GARDEN USE

Give them plenty of room, but use some punchy colors and contrasting forms nearby, such as crocosmia and, in part shade, sword fern.

SPECIES AND CULTIVARS

A. mollis—showy foliage waxy, glossy, deeply cut with wavy edges, leaves 24 in. long; flowers hooded, funnel-shaped, white and purple. Blooms early summer. 5 ft. x 3 ft. Zones 7–10. 'Hollard's Gold' foliage yellow-gold; 'Tasmanian Angel' foliage splashed white; 'Whitewater' similar, improved.

A. spinosus—spiny-tipped, wide, dark leaves like a thistle's; flowers two-toned mauve and white. Blooms summer. 5 ft. x 3 ft. Zones 5–9.

Achillea 'Terra Cotta'

Achillea

(yarrow) *Asteraceae*

Summer-blooming plant known for its flat-topped clusters of tiny flowers. Most have well-cut, ferny foliage.

CULTIVATION

Full sun and well-drained, even sandy soil. Cut back after flowers begin to fade to produce more flower stems, to remove unsightly finished flowers (light-colored cultivars fade to unattractive, dirty white), and to reduce reseeding. If grown in full sun, rarely needs staking, but rather than trying to stake it if it flops, provide nearby plants to hold it up and enjoy the new combination as flower stems weave into neighboring plants. Plants increase by underground shoots; divide or decrease by chopping through and digging out.

GARDEN USE

Excellent color and form to combine with golfball flowers of globe thistle (*Echinops*), round mounds of lavender, and spiky flowers of hebes. Brings butterflies, ladybugs, and other beneficial insects into the garden.

SPECIES

A. filipendulina—evergreen clumps of gray-green foliage; yellow, flat-topped heads of yellow flowers 5 in. across. Blooms early summer to early autumn. 4 ft. x 1.5 ft. Zones 3–9. 'Cloth of Gold' mustard yellow.
A. millefolium—mats of dark green leaves; magenta flower heads 4 in. across, fading with age. Blooms all summer. 24 in. x 24 in. Zones 3–9. 'Cerise Queen' dark pink; 'Moonshine' bright yellow; 'Paprika' red with white eye.
A. ptarmica—lance-shaped basal leaves; best selection 'The Pearl' or The Pearl group for clusters of small, double, white flowers. Blooms summer. 36 in. x 24 in. Zones 3–8.

CULTIVARS

(Mid- to late summer.)
'Apelblut'/'Appleblossom' pink flower clusters 3 in. wide, fading in summer heat; 36 in. x 24 in.; zones 4–8. 'Coronation Gold' evergreen, silver-gray foliage; golden 4-in.-wide flowers; 36 in. x 18 in.; zones 3–9. 'Freuerland'/'Fireland' brick red; 30 in. x 24 in. 'Summerwine' dark pink-red flower clusters 3 in. wide; 24 in. x 24 in.; zones 4–8. 'Terra Cotta' flowers fade to bronze.

Aconitum

(monkshood) *Ranunculaceae*

Tall plants with well-divided leaves and spires of hooded, dark, menacing-looking flowers.

CULTIVATION

Full sun or part shade. Provide regular water.

GARDEN USE

Wonderful rising out of low-growing plants; good dark flower colors, as well as flowers late in season. Grow with hardy geraniums, such as bright pink 'Patricia', and early flowering plants, such as wallflowers 'Julian Orchard' and 'Bowles Mauve'.

NOTES

All plant parts are poisonous.

SPECIES

A. carmichaelii—dark leathery leaves; indigo flowers. Blooms late summer. 5 ft. x 1 ft. Zones 3–8. 'Arendsii' rich blue.

CULTIVARS

'Bressingham Spire' deep violet; blooms midsummer to early fall; 36 in. x 12 in.; zones 3–8. 'Spark's Variety' deep violet, branched panicles; blooms mid- to late summer; 4 ft. x 1.5 ft.; zones 3–8. 'Stainless Steel' pale violet, branched; blooms early summer; 40 in. x 12 in.; zones 3–8. 'Ivorine' upright, bushy, ivory flowers; blooms late spring to early summer; 36 in. x 18 in.; zones 5–8.

Aconitum carmichaelii

Actaea

including Cimicifuga (baneberry, bugbane, snakeroot) *Ranunculaceae*

A genus of woodland plants that now includes the genus Cimicifuga, a taxonomic change that will take years to filter down to the garden level. Baneberries are short; Cimicifuga is tall and has slender, bottle-brush-like flower heads. Actaea is also known as doll's eyes, for its white berries with a dark spot.

CULTIVATION

Part shade and humusy soil; use a good mulch to help retain moisture. Prefers slightly acid soil, but will tolerate average pH.

GARDEN USE

Baneberries combine well with astilbes. Because woodland gardens usually have shrubs as middle story, be sure not to plant tall *Cimicifuga* under low-branching shrubs. White flowers, especially of late-blooming, former *Cimicifuga* selections, contribute greatly to shade gardens.

SPECIES

A. matsurmurae (syn. *Cimicifuga*) 'White Pearl'—clump-forming with well-divided leaves; bottlebrush white flowers atop stems. Blooms late summer. Shade to part shade. 4 ft. x 3 ft. Zones 4–9.

A. pachypoda—large, toothed leaves; white flowers followed by white berries in fall. Blooms spring. Shade to part shade. 24 in. x 18 in. Zones 3–9.

A. racemosa (syn. *Cimicifuga*; black cohosh)—clump forming with toothed, divided leaves; fluffy white flowers with unpleasant scent. Blooms midsummer. Shade to part shade. 4–7 ft. x 2 ft. Zones 3–8.

A. simplex (syn. *Cimicifuga*; snakeroot)—clump-forming with purple stems; white bottlebrush flowers. Blooms early to midfall. 4 ft. x 2 ft. Zones 4–8.

CULTIVARS

Atropurpurea group 'Black Negligee' 24 in. 'Brunette' brownish-purple leaves; 8-in. racemes of tiny, purple-tinged white flowers.

Actaea 'Brunette'

Agapanthus 'Prolific Blue'

Agapanthus

(lily-of-the-Nile) *Liliaceae*

Exotic-looking plants with strappy leaves growing from central point. Bare stems rising above foliage are topped with round or slightly drooping balls of funnel-shaped blue or lavender (sometimes white) flowers.

CULTIVATION

Full sun and well-drained soil; provide summer water. Although we long thought *Agapanthus* were not hardy in the maritime Northwest, they can be kept out in the garden in mild areas; give potted plants some protection in winter. Deciduous selections are more hardy here.

GARDEN USE

Fabulous in pots, where they provide a bit of the tropics to the patio or deck. Combine with some *Diascia* trailing over the edge of the pot, or use more than one pot where neighboring plants won't be smothered by *Agapanthus* foliage. Croscosmia and abutilons make colorful companions; use Kniphofia for similar form but different color.

SPECIES

A. africanus—evergreen; clumps of straplike leaves; flowers deep blue, 6–12 in. across. Blooms mid- to late summer. 36 in. x 18 in. Zones 9–10. 'Albus' white.
A. campanulatus—vigorously clumping, deciduous; narrow leaves; rounded flower heads 4–8 in. across. Blooms mid- to late summer. 4 ft. x 1.5 ft. Zones 7–10. Var. albidus white flowers.

CULTIVARS

Headbourne hybrids from deep violet to pale blue; deciduous. 'Lilliput' deciduous; deep blue; 16 in. x 16 in. 'Mood Indigo' deciduous; tubular, dark blue; 36 in. x 36 in. 'Peter Pan' evergreen; blue; 18 in. x 18 in. 'Storm Cloud' evergreen; dark blue; 24 in. x 24 in.

Agastache

(mint hyssop) *Lamiaceae*

A member of the mint family, with opposite leaves and small, tubular flowers in a variety of showy colors. Many have anise-mint scent to foliage.

CULTIVATION

Full sun and ordinary soil; several require no summer water. Cut back spent flowers for more blooms.

GARDEN USE

Great for summer color. More delicately blooming cultivars (OK—small flowers) can be used in pots or out in the garden with other sun-lovers, including asters, catmint, and sedum.

SPECIES

A. aurantiaca—erect, bushy, mint-scented spikes of orange-pink flowers. Blooms all summer. 18 in. x 24 in. Zones 7–10.

A. cana—erect, branched, with bubblegum/camphor smell; loose spikes of deep violet flowers. Blooms late summer to fall. 36 in. x 18 in. Zones 5–10.

A. foeniculum (anise hyssop)—erect, leafy, with anise scent; dense cylinders of violet flowers. Blooms mid-summer to early fall. Reseeds. 3–5 ft. x 1 ft. Zones 6–10.

A. rupestris—mounding, shrubby; grayish, aromatic leaves; narrow, tubular orange-and-pink flowers with prominent stamen. Blooms summer. No summer water. 24 in. x 24 in. Zones 5–8.

CULTIVARS

(Zones 6–9.)

'Apricot Sprite' peachy tint; 18 in. x 12 in. 'Apricot Sunrise' pale orange with purple tint; 36 in. x 24 in. 'Black Adder' lavender; 30 in. x 18 in. 'Coronado' orange-streaked yellow; 30 in. x 18 in. 'Firebird' loose spikes of copper orange; blooms midsummer to late fall; 24 in. x 24 in. 'Summer Breeze' peach-pink; blooms summer; 36 in. x 36 in. 'Tutti-frutti' raspberry red, loose spikes; 24 in. x 24 in.

Agastache 'Summer Breeze'

Alchemilla alpina

Alchemilla

(lady's mantle) *Rosaceae*

Mounding plants with round, lobed leaves. Flowers in sprays of chartreuse or light green.

CULTIVATION

Sun or part shade and ordinary soil. Slight reseeders, but rarely annoying; cut back *A. mollis* after it blooms to reduce this.

GARDEN USE

Grown more for foliage than flowers. Great plants to use for form. Common lady's mantle (*A. mollis*) catches drops of dew on hairy edges of its leaves; grow with ornamental onions for contrast in shape. Cultivars of *A. mollis* look little different from the species. Late-blooming plants such as *Rudbeckia* and tall asters add some color.

SPECIES

A. alpina (alpine lady's mantle)—may or may not be *A. conjuncta* in the trade; lobed leaves have silver hairs along edges and underneath, giving white variegated appearance; insignificant small flower sprays. Blooms summer. Reseeds politely in cracks of pavement. 5 in. x 20 in. Zones 3–8.

A. ellenbeckii—dark green, rounded, serrated leaves, red stem; evergreen mat. 2 in. x 12 in. Zones 6–8.

A. erythropoda—like a smaller *A. mollis*. Blooms summer. 12 in. x 8 in. Zones 3–8.

A. mollis (lady's mantle)—large dome of foliage; soft green, round leaves with slightly serrated edge; sprays of tiny chartreuse flowers. Blooms early summer. 24 in. x 30 in. Zones 4–8.

Allium

(ornamental onion) *Alliaceae*

Balls of flowers grow atop unclothed stems. Often, foliage dies back before or just as flowering begins in early to midsummer.

CULTIVATION

Full sun and well-drained soil a must for the best, most upright stems. Some reseeding.

GARDEN USE

Fabulous foil to low, mounding plant shapes such as hardy geranium. Also use later-blooming plants such as *Aster* 'Mönch'. Looks good popping out of ground-cover roses. Leave flower stems to age and turn to seedheads—they'll look good for several months.

NOTES

Because ornamental onions grow up rather than out, only height is listed here.

SPECIES

A. aflatunense—lilac flowers, 4-in. heads. 36 in. Zones 4–8.

A. cristophii (star of Persia)—short stalk; impressive lavender flower heads 8 in. across. 12 in.

A. giganteum—purple, 4-in. flower heads. 5–6 ft. Zones 6–10.

A. karataviense—broad leaves; pinkish flowers; short stems, so don't hide it under a bushel. 10 in. Zones 5–9. 'Ivory Queen' white.

A. moly (golden garlic)—sunshine-yellow, small enough to fit anywhere; 2-in. flower heads. Sun to a little light shade. 10 in. Zones 3–9. 'Jeanine' more yellow.

A. schubertii—like a fireworks show in your garden; 12-in. lavender flowers spaced out; beautiful in garden or a vase. 12–24 in. Zones 4–10.

A. schaerocephalon—crimson purple, 2 in. wide. 30 in. Zones 5–9.

CULTIVARS

'Firmament' 3-in. heads of lavender flowers with silver anthers; 24 in. 'Gladiator' purple, 6-in. heads; 5 ft. 'Globemaster' 8 in. across; 32 in.; zones 6–10. 'Lucy Ball' 3 in. across; 36 in.; zones 6–10. 'Purple Sensation' 3-in. heads; 36 in.; zones 4–9.

Allium cristophii

Alstroemeria, Pacific Sunset hybrid

Alstroemeria

(Peruvian lily) *Amaryllidaceae*

Colorful, tall, and aggressive. Various strains and cultivars for tubular flowers in shades of yellow, apricot-pink, and orange. The lance-shaped leaves are slightly twisted.

CULTIVATION

Full sun and well-drained soil. Spreads underground by means of rhizomelike tubers—to the annoyance of some gardeners; corral by means of a root barrier (sink several inches in the ground around the plant) or give it a bed to itself. Plants die back by late summer.

GARDEN USE

Once corralled, *Alstroemeria* is a good companion for late-summer flowers, such as *Rudbeckia*. The cut flowers are wonderful for indoor arrangements.

CULTIVARS

Pacific Sunset hybrids apricot, pink, and yellow mix; blooms summer; 24 in. x 36 in.; zones 7–10.

Amsonia

(blue star) *Apocynaceae*

Billowing mounds of narrow foliage. Clusters of blue flowers in early summer.

CULTIVATION

Sun and well-drained soil.

GARDEN USE

Pleasing sight with sky-blue flowers and soft texture. Grow with dark-foliaged sedum such as 'Bertram Anderson' and upright bronze sedge *Carex buchananii* 'Red Fox'.

SPECIES

A. hubrectii—green leaves turn into bright yellow fall foliage; panicles of sky-blue flowers. Blooms late spring. 3 ft. x 4 ft. Zones 5–9.

A. tabernaemontana (willow blue star)—narrow, light green leaves; dense rounded panicles of pale blue flowers. Blooms late spring. 24 in. x 18 in. Zones 3–8. Var. *salicifolia* leaves more narrow; flower heads more open.

Amsonia tabernaemontana

Anemone x hybrida

Anemone

(windflower) *Ranunculaceae*

Wide variety of perennials for woodland or open border, some low-growing with ferny foliage and daisylike flowers, others tall with lobed foliage and open flowers with broad petals. Flower color ranges from white to blue, or in shades of pink.

CULTIVATION

Light requirements according to species and cultivars, below; all prefer well-drained soil. Cultivars of hupehensis and x hybrida can take dry soil in part shade. Tall late anemones take cutting back hard rather well—sometimes the best way to keep them under control, because they tend to widen and widen and widen.

GARDEN USE

Early bloomers give gardens a boost; plant with small bulbs such as *Chionodoxa* and hardy geraniums that cover bare spots when anemones go dormant. Late-flowering anemones offer color and can be good fillers at back of borders.

SPECIES AND CULTIVARS

A. blanda—delicate, ferny foliage with dark stems that dies back by summer; white, blue, or pink flowers open on sunny days. Blooms early spring. Sun; grows from little knobs that look like pieces of wood. 6 in. x 6 in. Zones 4–8.

A. coronaria (poppy-flowered anemone)—bright-colored flowers in pink, purple, and red. Blooms spring. Sun; grows from knobby tubers. 18 in. x 6 in. Zones 8–10.

A. hupehensis—woody base; basal leaves and tall flower stems. Blooms late summer. Sun to part shade; grows from fibrous rootstock that suckers. 36 in. x 16 in. Zones 4–8. 'Prinz Heinrich'/ 'Prince Henry' dark pink.

A. x *hybrida*—similar to above and common in gardens; basal leaves serrated. Blooms late summer. Sun to part shade; tall flower stems flop in too much shade. 4–5 ft. x 2 ft. Zones 4–8. 'Honorine Jobert' white; 'Konigin Charloette'/'Queen Charlotte' semidouble, pink. 'Wirbelwind'/'Whirlwind' semidouble, white. 'September Charm' pale pink.

A. nemorosa (wood anemone)—ferny foliage dies back in summer; star-shaped white flowers. Blooms spring. Part shade. 6 in. x 12 in. Zones 4–8. 'Robinsoniana' light blue.

A. sylvestris (snowdrop anemone)—low plant; toothed, lobed leaves; white flowers. Blooms late spring. Sun to part shade; spreads by rhizomes. 12 in. x 12 in. Zones 3–8.

Angelica gigas

Angelica

(archangel) *Apiaceae*

Tall, architectural plants with hollow stems. Terminal flower heads like big fists. Member of the carrot family; encourages beneficial insects such as butterflies, ladybugs, and hover flies in the garden.

CULTIVATION

Full sun and well-drained soil. Short-lived; can act as a biennial if let go to seed—but flower heads are fabulous, so you may want to collect seeds and start over.

GARDEN USE

Height of Angelica is good foil for a low planting of *Bergenia* with its big leaves. The purple stems and fist-shaped flower heads of *A. gigas* contrast well with the late, bright color of *Rudbeckia*.

NOTES

Angelica seeds are used as a culinary flavoring; make your own liqueur.

SPECIES

A. archangelica—coarse green leaves; greenish flowers 10 in. across. Blooms early to midsummer. 6 ft. x 4 ft. Zones 4–9.

A. gigas—dark purple stems; plum-colored buds open to flower heads 5 in. across. Blooms late summer. 3–6 ft. x 4 ft. Zones 4–9.

Anthemis tinctoria

(golden marguerite) *Asteraceae*

Sunny, daisy flowers and ferny foliage. Long summer bloom time.

CULTIVATION

Full sun and well-drained soil. Zones 3–8.

GARDEN USE

Cultivar colors easy to combine with tall campanulas or dame's rocket. Late asters can carry on the show. 36 in. x 36 in.

CULTIVARS

'E. C. Buxton' lemon yellow. 'Moonlight' light yellow fading to white. 'Sauce Hollandaise' pale cream. 'Susanna Mitchell' cream with bright yellow center.

Anthemis 'Susanna Mitchell'

Aquilegia, Barlow hybrid

Aquilegia

(columbine) *Ranunculaceae*

A proliferation of hybrids and hybrid strains, which attests to how easily the plant crosses and changes (and how much we love spring). Tubular flowers, often with prominent spurs and sometimes two-toned. Foliage ferny and basal.

CULTIVATION

Full sun or light shade and well-drained soil. Soil moist in spring, but it helps if a good mulch is applied. Plants may seed around; cut back spent flowers if you don't want this.

GARDEN USE

Combine with other spring flowers such as Jacob's ladder (*Polemonium*) and the mourning widow geranium (*G. phaeum*). Later-flowering geraniums fill in columbine's empty spaces, when all that's left for the rest of summer is a rosette of leaves.

NOTES

Columbine sawfly is a spring pest that damages more in some years than others. Inspect foliage occasionally; if you see it eaten away, turn the leaf over to find a small green worm (the same color as the foliage) doing the damage. Wear gloves and dispatch manually. Leaf miners run roads through foliage; if you catch them in the act, do the same as for the sawfly.

SPECIES

A. caerulea—sky-blue flowers and straight spurs. Blooms late spring. 24 in. x 12 in. Zones 3–8.
A. canadensis—drooping, tubular scarlet flowers with lemon yellow petals peeking out. Blooms midspring to midsummer. 36 in. x 12 in. Zones 3–8. 'Corbett' light yellow.
A. chrysantha—yellow, long-spurred flowers on bushy plant. Blooms late spring. 36 in. x 24 in. Zones 3–8.
A. flabellata—short-spurred blue flowers; blue-green foliage. Blooms early summer. 12 in. x 6 in. Zones 4–9.
A. formosa—widespread orange sepals, yellow petals, and orange spurs. Blooms late spring to early summer. 36 in. x 18 in. Zones 4–8.
A. vulgaris (granny's bonnet)—violet and other shades of slightly spurred flowers. Blooms late spring to early summer. 36 in. x 18 in. Zones 3–8. Barlow hybrids (double): black Barlow, white Barlow.

CULTIVARS

Songbird series ('Bluebird', 'Chaffinch', 'Cardinal', 'Dove', 'Goldfinch') have widespread sepals that look like petals, with contrasting petals. McKana hybrids have long spurs.

Armeria

(thrift) *Plumbaginaceae*

Small clumps of plants with narrow, almost grassy foliage. Lollipop flowers, usually pink.

CULTIVATION

Full sun and well-drained soil. Deadhead to prolong flowering.

GARDEN USE

Gets lost easily but so cute that you should accommodate its small stature. Grow in rockeries, cracks in patio pavement, along the edge of a path. Grow at the base of tall ornamental onions or spiky veronica or alongside ornamental oregano at the top of a retaining wall.

SPECIES

A. juniperifolia (syn. *A. caespitosa*)—miniature stature; pink flowers. Blooms early summer. 3 in. x 6 in. 'Bevan's Variety' rose-pink.

A. maritima—tight, grassy clumps of foliage; pink flowers. Blooms early summer. 8 in. x 12 in. 'Alba' white. 'Bloodstone' dark red-pink. 'Dusseldorf Pride' rose pink. 'Rubrifolia' dark foliage. 'Victor Reiter' dwarf form.

Armeria maritima

Artemisia 'Powis Castle'

Artemisia

(wormwood) *Asteraceae*

Silver and gray foliage can be dissected and ferny or leafy. Shrubby; grows from a woody base or has a suckering habit. Flowers usually not the main show and are often barely noticeable.

CULTIVATION

Full sun and well-drained soil. Suckering species should be contained if you don't want your mixed bed turning into a monoculture. Plants with a woody base (such as 'Powis Castle') resent being cut back into wood more than once or twice and will die off. Trim back dead foliage, but it's better to replace them than try too strict a management.

GARDEN USE

Silver foliage looks fabulous with pinks and purples, so toadflax, phlox, and asters can take you through almost the entire gardening year.

SPECIES AND CULTIVARS

A. absinthium—coarse, shrubby; tiny gray flowers. Blooms summer. 36 in. x 24 in. Zones 4–8.

'Lambrook Silver' deeply divided silver leaves; refined look; 30 in. x 30 in. 'Powis Castle' feathery silver foliage; 24 in. x 36 in. Zones 7–9.

A. lactiflora—an exception to the silver-foliage/poor flower rule, this has sprays of white, fragrant flowers. Blooms late summer. Sun to part shade. 5 ft. x 2 ft. Zones 5–8. 'Guizhou' (Guizhou group) purple-flushed stems; white flowers.

A. ludoviciana—lance-shaped gray leaves on white stems; highly ornamental, looks great in cut-flower arrangements. Blooms summer. Runs underground. 36 in. x 24 in. Zones 4–9. 'Silver Queen' 30 in. x 24 in. 'Valerie Finnis' 24 in. x 24 in.

A. schmidtiana 'Silver Mound'—soft, silver mound of foliage (go ahead, pat it). Blooms summer. Looks best if cut back hard in June; otherwise it opens up in the middle. 12 in. x 18 in. Zones 5–8.

A. stelleriana 'Silver Brocade'—dense, silver-felted cut leaves; yellow daisy flowers nothing remarkable; similar to but not the same as another plant called dusty miller (Senecio). Blooms summer. 6 in. x 18 in. Zones 3–8.

Arum italicum

Arum italicum
(arum) *Araceae*

Arrow-shaped leaves emerge in winter and die back in summer, leaving a surprise—upright stems with red fruit, looking as if someone had stuck the end in a bowl of berries. Flowers unnoticeable.

CULTIVATION

A great plant for shade and dry soil.

GARDEN USE

Grow with *Epimedium*, hardy cyclamen, wood aster, and wild ginger. Zones 6–9.

CULTIVARS

'Pictum', 'Marmoratum', or var. *italicum*—leaves marked with silver or cream; blooms late spring; 12 in. x 12 in.

Aruncus

(goatsbeard) *Rosaceae*

*Divided foliage. Tiny white flowers in clusters; frothy
white flowers give this genus its other common name:
false astilbe.*

CULTIVATION

Part shade in well-drained soil. Tall goatsbeard takes
our normally dry summers but does end up looking a
little on the dry side. Regular water for other species.

GARDEN USE

A. aethusifolius grows well with lungwort and Jacob's
ladder, taller species with hostas and foamflower.

SPECIES

A. aethusifolius—more delicate than its cousin; ferny
green foliage turns red in fall before dying back;
crooked stems of flowers reminiscent of gooseneck
loosestrife. Blooms early to midsummer. Sun to part
shade. 16 in. x 16 in. Zones 4–8.

A. dioicus—the more common goatsbeard; bolder
foliage; sprays of flowers on taller stems. Blooms
midsummer. Sun to part shade. Dry summer soil
OK. 6 ft. x 4 ft. Zones 3–8.

Aruncus aethusifolius

Asarum caudatum

Asarum caudatum

(wild ginger) *Aristolochiaceae*

Spreading native ground cover with heart-shaped, dark evergreen leaves. Hard-to-see flowers are held under leaves in spring.

CULTIVATION

Shade with no supplemental water. Zones 4–8.

GARDEN USE

Good in the native plant garden under conifers and shrubs.

Asclepias

(butterfly weed, milkweed) *Asclepidaceae*

Lance-shaped leaves with clusters of bright flowers atop stems. Much loved by aforesaid insects. Blooms in summer.

CULTIVATION

Full sun. Water according to species. Zones 4–9.

GARDEN USE

Grow near early spring bulbs, where butterfly weed will help cover dying foliage.

SPECIES

A. incarnata (swamp milkweed)—pink flowers; wet soil; 4 ft. x 2 ft. 'Cinderella' white flowers from red bracts. Regular water.

A. speciosa—wide leaves; purple-pink flowers; spreads by rhizomes; 30 in. x 24 in. Dry summer soil OK.

A. tuberosa—spiraled, lance-shaped leaves; bright orange or yellow flowers; ornamental seedpods; takes dry soil. 36 in. x 12 in. Regular water.

Asclepias tuberosa 'Gay Butterflies'

Aster

(aster) *Asteraceae*

Wide range of perennials that bloom throughout summer and into fall. Daisylike flowers range from small to wide. Plants tall and thin or short and stout. Leaves are oblong to lance-shaped.

CULTIVATION

Full sun and well-drained soil. Regular water except as noted below. Tall selections may lean over as they bloom; stake or provide supporting neighbors. Problems with powdery mildew can be ameliorated by providing good air circulation or just cutting down when flowering is finished.

GARDEN USE

No end to the combinations. Because many bloom late in the season, they can take over for early-flowering perennials. 'Mönch' and 'Wunder von Stafa' begin in midsummer and bloom for weeks and weeks; combine these with contrasting flowers and foliage of crocosmia and tall goldenrods. Tall asters benefit from some dressing at their base—for example, lady's mantle, *Geranium renardii*, and sun-loving coral bells.

SPECIES AND CULTIVARS

A. cordifolius (blue wood aster)—erect stems; sprays of 1/2-in. blue flowers. Blooms late summer to early autumn. Light shade. 5 ft. x 2 ft. Zones 5–8. 'Little Carlow' to 4 ft.

A. divaricatus (wood aster)—a lax plant, it weaves around; small, starry white flowers brighten up tired plantings. Blooms midsummer to midautumn. An exception to the rule: likes shade to part shade; doesn't mind dry soil in summer. 24 in. x 24 in. Zones 4–8.

A. ericoides var. *prostratus* 'Snow Flurry'—forms a mat; short, dense leaves; single white flowers. Blooms late summer. 36 in. x 12 in. Zones 5–8.

A. x *frikartii* 'Mönch', 'Wunder von Stafa'—lovely, easy plants, both in shades of lavender-blue. Bloom for weeks and weeks midsummer to early autumn. 28 in. x 16 in. Zones 5–8. 'Flora's Delight' lilac; 20 in. x 24 in.

A. lateriflorus—tall and lanky; sprays of tiny flowers. Blooms midsummer. 4 ft. x 1 ft. Zones 4–8. 'Horizontalis' 24 in. x 12 in. 'Lady in Black', 'Prince' dark foliage and pink buds, giving flower sprays a raspberry color.

A. novae-angliae (New England aster)—blooms late summer to midautumn. 5 ft. x 2 ft. Zones 4–8. 'Andenken an Alma Potschke' bright pink. 'Harrington's Pink' pale pink. 'Purple Dome' purple; 24 in. x 36 in.

A. novae-belgii (syn. *A. dumosus*; Michaelmas daisy)—blooms late summer to midautumn. 4 ft. x 3 ft. Zones 4–8. Shorter selections: 'Alice Haslam' deep, double pink; 10 in. 'Patricia Ballard' double pink; 30 in. 'Prof. Anton Kippenberg' lavender-blue; to 12 in. 'Wood's Pink' deep pink; to 15 in.

Aster 'Flora's Delight'

Astilbe 'Elizabeth Bloom'

Astilbe

(astilbe) *Saxifragaceae*

A group of plants known for a delicate appearance. Ferny, often bronze or red foliage. Frothy stems of pink, red, or white flowers.

CULTIVATION

Sun to shade (most grow best in shade, some do well in sun) and moist soil. All prefer regular water; if they don't get it, they'll look terrible. *A. chinensis* cultivars may take dry shade; want to try it?

GARDEN USE

Grow with like-minded plants—*Rodgersia* and *Brunnera*, for example. Astilbes can be sequenced for a long flowering time by picking early-, mid-, and late-season bloomers.

SPECIES AND CULTIVARS

A. x *arendsii*—blooms early summer. Part shade. 36 in. x 36 in. Zones 4–8. Early-season cultivars: 'Deutschland' white. 'Fanal' crimson. 'Rheinland' bright pink. Midseason cultivars: 'Amethyst' lilac. 'Brautscheieler'/'Bridal Veil' white. 'Bressingham Beauty' pink. 'Elizabeth Bloom' pink. Late-season cultivars: 'Cattyela' rose-pink. 'Snowdrift' white.

A. chinensis var. *pumila*—dark green, deeply divided leaves; pink or violet flowers. Blooms late summer. Sun to part shade. 24 in. x 24 in. Zones 4–8. 'Finale' to 18 in. 'Pumila' to 12 in.

A. simplicifolia hybrids include 'Sprite'—pink. Blooms late summer. 12 in. x 8 in. Zones 4–8.

Astrantia

(masterwort) *Apiaceae*

Clumping plant that increases by underground runners. Coarsely lobed and slightly serrated leaves. Small pincushion of true flowers surrounded by pointed "petals" that are really leafy structures called bracts. One-inch flowers white with green vein on bracts, occasionally with pink tinge; flowers rise above foliage on thin stalks.

CULTIVATION

Full sun to a little light shade and well-drained, evenly moist soil.

GARDEN USE

Grow with contrasting colors, such as crocosmia, or stay with cool tones and combine with Stoke's aster and *Clematis* x *durandii*.

SPECIES AND CULTIVARS

A. major—mostly known for its cultivars, which have stretched the range of flower color. Bloom early to midsummer. 12–36 in. x 18 in. Zones 4–8. 'Claret' pink. 'Hadspen Blood' dark red. Rubra' crimson. 'Ruby Cloud' pink. 'Ruby Wedding' ruby-red with dark stems. Subsp. *involucrata* 'Shaggy' larger flowers with white, long bracts tinged green.

A. maxima—leaves divided into three lobes; flowers white with pink tint. Blooms summer. 24 in. x 12 in. Zones 4–8.

Astrantia 'Hadspen Blood'

Athyrium niponicum var. *pictum*

Athyrium niponicum

var. *pictum* (Japanese painted fern) *Athyriaceae*

Fronds emerge dark maroon and expand to reveal a pewter background with dark markings along the midrib.

CULTIVATION

Shade to part shade and evenly moist soil.

GARDEN USE

Fabulous for color and texture in shade. Combine with early-blooming lungworts. Pick up dark tones by using purple-flowering hellebores. Continue pewter theme with small *Hosta* 'Blue Cadet'.

CULTIVARS

Sometimes listed as 'Pictum'—12 in. x 24 in. Zones 5–8.

Ballota pseudodictamnus

(ballota) *Lamiaceae*

Small, shrubby plants with soft, silver-green, round leaves and tubular flowers.

CULTIVATION

Full sun and well-drained soil. Cut back hard in early spring for a cleaner look. Zones 7–10.

GARDEN USE

Blooms midsummer, although blooms can be sheared back for more foliage. Grow with other sun-lovers. Good with lamb's ears (Stachys), lavender, and sun roses. 24 in. x 24 in.

Ballota pseudodictamnus

Baptisia australis

Baptisia

(false indigo) *Fabaceae*

Large herbaceous plant with a mounding, not totally upright appearance. Leaves divided into three oblong leaflets held close to the stem. Pealike flowers appear on long stalks at tops of stems in early summer.

CULTIVATION

Full sun and well-drained, lean soil; a good plant for one of those hot spots with sandy soil. Give it some room or sturdy neighbors, as it will probably flop about. Seedheads are ornamental black, flat peapods, but once the plant sets seed it ceases to bloom; prolong flowering by deadheading, or enjoy the fruit of no action. Cut down in winter. Zones 4–9.

GARDEN USE

Grow with upright *Boltonia* or use in the background with penstemon, lavender, and bronze sedge. An easy-care perennial for a lovely, sunny look.

SPECIES

B. alba—leaflets divided into three; whiteflowers. 2–4 ft. x 2 ft.
B. australis—sturdy, upright plant with grayish foliage; indigo flowers. 5 ft. x 2 ft.

CULTIVARS

'Purple Smoke' dusky purple flowers; 4 ft. x 3 ft.
'Solar Flare' sunny yellow flowers; 4 ft. x 4 ft.

Begonia grandis

(hardy begonia) *Begoniaceae*

Hard to believe there's a begonia—known mostly as a houseplant or summer annual—that will live over in the garden, but B. grandis *is a hardy, herbaceous perennial. Leaves infused with purple-red that shows up well if you situate plant so afternoon sun is behind it. Small sprays of pink or white flowers in summer.*

CULTIVATION

Shade to part shade and moist, well-drained soil with a good mulch.

GARDEN USE

Makes a dramatic statement in the shady woodland garden. Combine with ferns, *Rodgersia*, *Omphalodes*, and other shady characters.

SPECIES AND CULTIVARS

B. *g.* subsp. *evansiana*—pink flowers. Blooms summer. 24 in. x 18 in. Zones 6–9. 'Alba' white. 'Heron's Pirouette' deep pink.

Begonia grandis

Bergenia 'Bressingham Ruby'

Bergenia

(elephant ears, pig squeak) *Saxifragaceae*

Old-fashioned, evergreen perennial that has enjoyed a renaissance of late with an influx of selections that show colorful foliage in winter. Big, glossy leaves grow in rosettes. Bare, dark stems rise above foliage in late winter carrying clusters of small pink flowers.

CULTIVATION

Sun or part shade and well-drained soil; foliage color develops best in full sun. Cut back winter-worn leaves in early spring or before flowers appear. Plants increase by offsets; divisions are easily made by slicing through roots that connect one rosette with another.

GARDEN USE

Mostly grown in rockeries and at edges of paths where they can be seen well. Grow with evergreen bronze sedges, summer-flowering penstemon, and variegated sedges.

NOTES

Root weevils can create a well-pinked edge on the leaves; use mechanical means (go hunting at night with a flashlight) to control.

SPECIES

B. cordifolia—bronze-tinged, heart-shaped leaves; pink or white flowers. Blooms late winter to early spring. 18 in. x 24 in. Zones 4–8. 'Winterglut' dark foliage; deep pink flowers; 24 in. x 30 in.

CULTIVARS

Bloom late winter to early spring; 18 in. x 24 in. unless otherwise stated; zones 2–9. 'Abendglocken'/'Evening Bells' dark winter foliage; deep pink flowers. 'Abendglut'/'Evening Glow' maroon foliage; magenta flowers. 'Apple Blossom' dark stems; pale pink flowers. 'Baby Doll' 8 in. x 18 in. 'Bressingham Ruby' dark foliage, especially in winter. 'Ruby Elf' to 6 in.

Boltonia asteroides 'Snowbank'

Boltonia asteroides

(thousand-flowered aster) *Asteraceae*

*Tall, airy plants with mostly basal foliage and large
clusters of tiny, white daisy flowers.*

CULTIVATION
Sun and well-drained soil.

GARDEN USE
Blooms late summer into fall, providing stature in
floral clouds. Use with shrub or ground-cover roses
and hardy geraniums. 6 ft. x 3 ft. Zones 4–9.

CULTIVARS
'Pink Beauty' light pink flowers. 'Snowbank' to 5 ft.
Var. latisquama 'Nana' 4 ft. x 3 ft.

Brunnera macrophylla

(Siberian bugloss) *Boraginaceae*

Sprays of blue flowers atop large, heart-shaped leaves.

CULTIVATION

Shade to part shade and well-drained, moist soil. Zones 3–8.

GARDEN USE

Blooms early spring. A comfortable woodland companion. Cultivars that brighten up the foliage make for a longer show. Combine with more blues—*Omphalodes* and lungworts—and short, arching stems of starry false Solomon's seal (*Smilacina stellata*). 18 in. x 24 in.

CULTIVARS

'Dawson's White' (also listed as 'Variegata') foliage splashed creamy white. 'Hadspen Cream' foliage has wide creamy border. 'Jack Frost' foliage silver with green veins. 'Looking Glass' silver cast to leaves.

Brunnera 'Dawson's White'

Calamagrostis x acutiflora 'Overdam"

Calamagrostis x acutiflora

(feather reed grass) *Poaceae*

Herbaceous, upright grasses with two exceptional cultivars for the garden: 'Overdam' fine white variegation on the narrow leaves; 36 in. x 24 in. 'Karl Foerster' green grassy base and tall, narrow bundle of flower stems; 5 ft. x 1.5 ft. Both age to warm straw color.

CULTIVATION

Sun to part shade. Cut down in late winter. Zones 5–9.

GARDEN USE

Provides a vertical note among mounding perennials including geraniums and *Rudbeckia*.

Campanula

(bellflower, harebell) *Campanulaceae*

Large genus with plants of varying size and shape, but all with telltale bell-shaped flowers, usually in white or lavender-blue, occasionally pink. Foliage is variable depending on the species—some have small, round, scalloped leaves while others have long narrow leaves.

CULTIVATION

Sun and well-drained soil, except as noted below. Tall campanulas may flop if not given enough sun; pinching back tall plants in early spring can keep them from flopping. Some mat-forming species like to spread and spread.

GARDEN USE

There are campanulas to fit into any garden in sun or shade. Mat-forming species make good rock-garden or edging plants; taller varieties grow well with smaller, mounding plants. Spring-flowering campanulas should be combined with summer-blooming plants so that the garden doesn't have a one-shot look. Many taller campanulas make good cut flowers.

SPECIES AND CULTIVARS

C. carpatica (Carpathian bellflower)—mat-forming; toothed, egg-shaped leaves; cup-shaped, upright blue flowers up to 2 in. across. Blooms early to mid-summer and longer. 9 in. x 24 in. Zones 4–8. 'Blaue Clips' blue flowers; 6–8 in. x 24 in. 'Weisse Clips' white flowers.

C. cochlearifolia (fairy's thimble)—mat-forming; small, heart-shaped leaves; loads of tiny blue flowers on thin stems held above foliage. Blooms summer. Sun to part shade. 3 in. x 12 in. Zones 5–8. 'Elizabeth Oliver' double.

C. garganica (Adriatic bellflower)—mat-forming; roughly serrated, heart-shaped leaves; pale blue, upward-facing, star-shaped flowers. Blooms summer. Sun to part shade; would like to take over the garden. 2 in. x 12 in. Zones 4–8. 'Dickson's Gold' yellow foliage; less aggressive, needs more shade.

C. glomerata (clustered bellflower)—upright; long, pointed leaves on dark red stems; blue, upright flowers in dome-shaped clusters at tops of stems. Blooms all summer. Spreads underground; especially aggressive in wet soil. 18 in. x indef. Zones 3–8.

C. lactiflora (milky bellfower)—tall stems with long, pointed leaves topped with big, heavy clusters of flowers. Blooms early to late summer; an impressive show. Needs staking, or let it lean into neighbors.

Campanula poscharskyana

5 ft. x 2 ft. Zones 5–8. 'Loddon Anna' pink flowers. 'Pritchard's Variety' dark blue; 30 in. x 24 in.

C. persicifolia (peach-leaved bellflower)—upright, reedy; narrow leaves; cup-shaped, 2-in., outward-facing white or lavender-blue flowers in loose clusters on top half of stems. Blooms early to midsummer. Sun to part shade. 36 in. x 12 in. Zones 3–8. 'Alba' white flowers.

C. portenschlagiana (Dalmatian bellflower)—mat-forming, spreading; small, kidney-shaped leaves; open, upright, star-shaped lavender flowers in profusion; good for crevices and cracks in pavement. Blooms mid- to late summer. Sun to part shade. 6 in. x 20 in. Zones 4–8. 'Birch Hybrid' purple-blue flowers; less aggressive.

C. poscharskyana (Serbian bellflower)—trailing, aggressively spreading; toothed leaves; starry, lavender flowers. Blooms summer to autumn. 6 in. x 24 in. Zones 4–8.

C. punctata (spotted bellflower)—upright with stems slightly leaning; basal leaves large, oblong; 2-in., closed and drooping white or light pink (spotted inside) flowers in loose clusters atop stems. Blooms early summer. Sun to part shade. 12 in. x 16 in. Zones 4–8. 'Cherry Bells' deep pink flowers. 'Hot Lips' pink with dark rose inside. 'Wedding Bells' hose-in-hose (double) white flowers.

C. rotundifolia—mat-forming; heart-shaped basal leaves and taller flowering stems; 1-in. blue flowers, nodding. Blooms early to late summer. 12 in. x 18 in. Zones 5–8.

C. trachelium (nettle-leaved bellflower, bats-in-the-belfry)—leafy base; pointed leaves long and wide with jagged edges; flowers lavender-blue or white. Blooms mid- to late summer. Sun to part shade. 36 in. x 12 in. Zones 5–8. 'Bernice' double violet-blue.

Carex

(sedge) *Cyperaceae*

Grasslike plant growing from a central clump. Leaves narrow and grassy or up to 1 in. wide. Flower stems not showy and rarely rise far above foliage, so bloom times not given below.

CULTIVATION

Sun or shade, according to species below, and moist, well-drained soil. Some shade-loving selections grow well without supplemental water if mulched. Evergreen sedges do not need cutting down every year, but it helps to groom those with grassy leaves. If winter damage has left ugly foliage, cut the whole plant back. Some stay put and others spread underground.

GARDEN USE

Use to great effect in the mixed garden. Generally, sedges with bronze or warm-tinted foliage do well in sun; combine these with roses or, for low-water situations, lavender, sun roses, verbascum, globe thistle, and sea holly. Sedges for shade contrast well with the big leaves of hostas and can cover up the base of clematis well.

SPECIES AND CULTIVARS

C. buchananii (bronze sedge)—evergreen, grassy, upright, with curlicue tips. Sun to part shade. 30 in. x 36 in. Zones 6–9. 'Red Fox' may be more rusty red, but it's too close to say for sure; 'Viridis' foliage silver-green, more lax.

C. comans (New Zealand hair sedge)—evergreen, lax. Sun to part shade. 15 in. x 36 in. Zones 7–9. 'Bronze' much like *C. flagellifera*. 'Frosty Curls' blue-green foliage.

C. conica 'Snowline' (syn. 'Marginata', 'Hime-kan suge')—evergreen; white edges to narrow leaves; brings light to dark corners. Shade. 6 in. x 10 in. Zones 5–9.

C. elata 'Aurea' ('Bowles Golden')—herbaceous, bright golden yellow foliage in a lax mound. Shade. 18 in. x 30 in. Zones 5–8.

C. flagellifera—mop of weeping brown evergreen foliage. Sun to part shade. 15 in. x 36 in. Zones 7–9.

C. morrowii 'Variegata'—broad white stripe down middle of half-inch-wide leaves. Part shade. 20 in. x 24 in. Zones 7–9. 'Ice Dance' creamy edges.

C. oshimensis 'Evergold'—narrow green leaves with a center gold stripe. Sun to part shade. 1 ft. x 2 ft.

C. testacea (orange sedge)—evergreen, grassy; warm orange tints, especially in winter. Sun to part shade. 36 in. x 24 in. Zones 8–9.

Carex flagellifera

Catananche caerulea

Catananche caerulea

(cupid's dart) *Asteraceae*

Stiff, ridged stems and few leaves. Impressive rayed violet flowers serrated at the tips.

CULTIVATION

Sun and well-drained soil. Zones 3–8.

GARDEN USE

Blooms for long period midsummer. Grow with catmint and late asters such as pink 'Andenken an Alma Potschke'. 36 in. x 12 in.

Centranthus ruber

Centranthus ruber

(Jupiter's beard) *Valerianaceae*

Succulent, glossy foliage. Domes of tiny magenta
flowers.

CULTIVATION

Sun to part shade and poor soil. Doesn't need summer water. Old-fashioned plant that self-sows but not to excess; ultimate easy-care perennial. Likes to reseed in difficult-to-reach places, such as tops of walls. Cut back spent flowers for more blooms. Zones 5–8.

GARDEN USE

Blooms late spring to late summer. Combine with campanula, lady's mantle, and sedum. 36 in. x 36 in.

Centaurea

(bachelor's button; cornflowers) *Asteraceae*

In spring, wide, ruffly flowers with a center disc from ornamental buds.

CULTIVATION

Full sun and well-drained soil. Cut back after flowering for another round. Zones 3–9.

GARDEN USE

Grow in the dry garden with sun roses and lavender.

SPECIES AND CULTIVARS

C. hypoleuca 'John Coutts'—leaves tri-lobed; mauve-pink flowers bloom spring and fall. 24 in. x 18 in.
C. montana 'Black Sprite'—long, lance-shaped leaves mostly basal; 2-in.-wide spidery, purple-black flowers. 24 in. x 24 in.

Centaurea montana

Cephalaria gigantea

Cephalaria gigantea
(giant scabiosa) *Dipsacaceae*

Impressive, though not bulky; grows from clump of large, divided, basal foliage. Almost leafless stems rise high above almost everything else and bloom with creamy yellow, pincushion flowers.

CULTIVATION
Sun and well-drained soil. Self-sows. Zones 3–8.

GARDEN USE
Blooms summer. Can grow with just about anything except small delicate plants that the lower foliage would smother. 8 ft. x 2 ft.

Chaerophyllum hirsutum 'Roseum'

(pink cow parsley) *Apiaceae*

A frothy delight with divided foliage like astilbe and flat-topped clusters of tiny pink flowers. Blooms spring.

CULTIVATION

Sun and well-drained soil. Regular water. Zones 6–8.

GARDEN USE

For the mixed border with hardy geraniums, veronicas, and campanulas. Let asters take over the show later. 24 in. x 24 in.

Cimicifuga—see *Actaea*

Chaerophyllum hirsutum 'Roseum'

Clematis recta 'Purpurea'

Clematis

(clematis) *Ranunculaceae*

Mostly known for woody vines, a few act more as herbaceous perennials. Leaves divided. Flowers variously drooping or open.

CULTIVATION

Full sun to part shade and well-drained soil; provide a good mulch. Cut back to the ground in winter. Provide a cage for support, or grow with supporting plants.

GARDEN USE

Even upright growers look best growing in the middle of sturdy companions, such as small conifers or ground-cover roses.

SPECIES

C. x *durandii*—lax, twining; open, indigo flowers with a cluster of white stamen. Blooms mid- to late summer over long period. 3–6 ft. x 3 ft. Zones 5–9.

C. integrifolia—upright to sprawling; slightly nodding flowers with four reflexed petals and cluster of white stamen. Blooms midsummer to early fall. 24 in. x 24 in. Zones 3–8.

C. recta 'Purpurea'—leaves emerge deep purple but fade slightly as summer progresses; at tops of stems, 1-in., lightly fragrant, four-petaled white flowers are reflexed with prominent cluster of white stamen. Blooms midsummer to autumn. 3–6 ft. x 2.5 ft. Zones 3–8.

CULTIVAR

'Alionushka' lax habit, grow through a cage or sturdy shrub; large, nodding, pink flowers, a sweet look; blooms summer to fall; 4 ft.; zones 5–9.

Convallaria majalis

Convallaria majalis

(lily-of-the-valley) *Liliaceae*

Broad leaves. Short stems of nodding, fragrant white flowers.

CULTIVATION

Shade to part shade and humusy, well-drained soil. Zones 2–7.

GARDEN USE

Blooms spring. Spreading ground cover. 9 in. x indef.

CULTIVARS

'Albostriata' yellow stripes down leaf veins. Var. *rosea* pink flowers.

Coreopsis

(tickseed) *Asteraceae*

Broad range of summer plants with yellow or pink daisylike flowers often serrated at tips. Foliage leafy or ferny.

CULTIVATION

Full sun and well-drained soil. Deadheading encourages more flowers. Plants easy to divide in fall or spring, but delicate-looking, dark-leaved foliage of cultivars such as 'Moonbeam' often hard to spot as they come out of the ground.

GARDEN USE

Grow with contrasting forms, such as small New Zealand flax 'Jack Spratt' or upright bronze sedge *Carex buchananii*. Weaving with plants such as Geranium 'Ann Folkard' or *Potentilla* 'Miss Willmott' can create new scenes during blooming season.

SPECIES AND CULTIVARS

C. rosea—delicate, soft, needle-shaped leaves; 1-in., single pink flowers. Blooms summer to early fall. 24 in. x 12 in. Zones 4–8. 'American Dream' rose

Coreopsis rosea 'Sweet Dreams'

pink; to 10 in. x 24 in. 'Sweet Dreams' bicolor white with raspberry petal base.

C. verticillata (threadleaf coreopsis)—slightly rhizotomous; green, ferny foliage; 1-in. single, yellow flowers with dark center. Blooms late spring to early fall. 24 in. x 18 in. Zones 4–9. 'Crème Brûlée' canary yellow. 'Moonbeam' pale yellow; dark foliage; 18 in. x 18 in. 'Zagreb' center disk dark yellow.

Corydalis
(fumewort, fumitory) *Fumariaceae*

Assortment of herbaceous plants that look similar but take different culture. In general, short and mounding; foliage finely divided and ferny looking. Small and spurred blue or yellow flowers, some fragrant.

CULTIVATION
Shade, except as noted below, and well-drained, evenly moist soil.

GARDEN USE
Blue corydalis looks lovely in the woodland garden, where its delicate foliage provides great textures. Grow with spring bulbs, lungworts, and *Brunnera*. *C. lutea* is tougher and can hold its own against robust plants such as Mrs. Robb's spurge (*Euphorbium amygdalordes* var. *robbiae*).

NOTES
Protect delicate blue corydalis from slugs.

SPECIES

C. cheilanthifolia (fernleaf corydalis)—evergreen; rosette with leaves that grow like long fern fronds from center of plant; yellow flowers. Blooms spring to early summer. Tuberous. 12 in. x 10 in. Zones 5–8.

C. flexuosa—divided foliage; blue flowers held on stems above leaves. Blooms early spring to early summer. Part shade. 12 in. x 8 in. Zones 6–8. 'Blue Panda' pale blue. 'China Blue' sky blue. 'Golden Panda' yellow foliage; blue leaves. 'Pere David' bright blue; spreading. 'Purple Leaf' bronze foliage; spreading.

C. lutea—almost evergreen; clumps of ferny foliage; practically continuous yellow flowers. Blooms mid-spring to fall. Sun to shade to part shade; bane of some gardeners for its reseeding ability, but truly easy-care and excellent for dry shade. 16 in. x 24 in. Zones 5–8.

CULTIVAR

'Blackberry Wine' wine-purple flowers; blooms mid-spring to summer; sun to part shade; 10 in. x 24 in.; zones 5–8.

Corydalis lutea

Crambe maritima

Crambe

(sea kale) *Brassicaceae*

Bold plant with large leaves that give great effect.
Blooms in sprays of white flowers.

CULTIVATION

Sun and well-drained soil.

GARDEN USE

Grow close-in with other plants, as sometimes foliage looks rather battered toward end of summer. *Phygelius*, *Rudbeckia*, and *Phormium* can all help.

NOTES

Bait for slugs with a nontoxic, phosphorus-based control.

SPECIES

C. cordifolia (giant sea kale)—large, green basal leaves 24 in. wide wrinkled with coarsely toothed edges; bare flower stems rise above, with huge clusters of small white, fragrant flowers. Blooms midsummer. 8 ft. x 5 ft. Zones 6–9.

C. maritima (sea kale)—lobed, gray-green leaves to 24 in. long; sprays of small, white, fragrant flowers. Blooms early summer. 30 in. x 24 in. Zones 6–9.

Crocosmia

(montbretia) *Iridaceae*

Strappy, irislike foliage. Midsummer flowers on slightly arching stems, opening gradually from base to tip.

Crocosmia 'Lucifer'

CULTIVATION

Sun and well-drained soil. Available as corms in fall or potted starting in spring. Garden-variety crocosmia is orange and rambunctious—it likes to spread and spread; dig out and toss oldest parts of the clump to keep it from taking over. Named cultivars increase more slowly.

GARDEN USE

Combines well in the garden with asters, lady's mantle, and phlox. Good cut flowers, but the seedheads, looking like beads on an arched stem, are longer-lasting.

CULTIVATORS

Bloom early to midsummer; 36 in. x 36 in.; zones 6–9. 'Bright Eyes' orange with red eye. 'Burnt Umber' red foliage; red-bronze flowers. 'Emberglow' deep scarlet; to 4 ft. 'Jenny Bloom' sunny yellow. 'Lucifer' deep scarlet; to 4 ft. 'Star of the East' deep gold with dark petal spot at base. *C.* x *crocosmiiflora* 'Emily McKenzie' two-toned, orange with dark petal base. 'Norwich Canary' daffodil yellow with orange reverse. 'Solfaterre' foliage tinted slightly bronze; flowers lemon yellow.

Cyclamen

Cyclamen

(cyclamen) *Primulaceae*

*Hardy cyclamen look just like the potted cylamen
at the grocery store, only in miniature. They appear
when we need them the most—from late summer
through winter—growing from corms and carpeting
the ground with pewter-patterned leaves. Flowers
in white and shades of lavender and pink look as if
they have been caught in a windstorm. Slowly increasing year by year. Two species: C. coum—leaves
more rounded; flowers in winter; C. hederifolium—
heart-shaped leaves; flowers in autumn.*

CULTIVATION

Shade. Summer-dry soil OK. Zones 5–9.

GARDEN USE

Easy care and essential components of Northwest
autumn and winter gardens. Plant them under
deciduous shrubs, such as witch hazel, or along a
well-traveled path. By late spring they disappear,
only to show their leaves again at the end of August.
4 in. x 12 in.

Delphinium

(delphinium) *Ranunculaceae*

Dramatic, tall, and elegant; the sight of a stand can cause gardeners to swoon. Flowers range from pale blue to deepest indigo (with occasional white or pink). Single stems, sometimes branching, have leafy, divided leaves halfway up and flowers covering the rest. How many cultivars are there? How many stars are there in the sky? Every real or imagined shade of blue has been slapped with a cultivar name, but they can mostly be grouped into named strains.

CULTIVATION

Full sun and rich, well-drained soil kept moderately moist. Cut back and fertilize for more flowers (one of the few perennials that needs more fertilizer than a mulch can provide). More plant stakes are bought for delphiniums than probably any other plant, because the taller they are, the more likely they are to fall. Good air circulation helps keep powdery mildew at bay.

GARDEN USE

Range of blues in summer looks good with just about anything—orange and yellow crocosmia and

Delphinium 'Guinevere'

early asters, for instance. Position tallest plants toward the back, but within reach if you want to fuss with reblooming.

SPECIES AND CULTIVARS

D. elatum—leaves divided like a hand; blue flowers. Blooms summer. 4–6 ft. x 2 ft. Zones 3–8. A parent of many hybrids and strains: Belladonna group many-branched; airy blue flowers; 5 ft. x 2 ft. Black Knight group large, dark blue flowers with black eye; 5 ft. English hybrids sturdy stems; vigorous repeat bloomers; white with dark eye, deep blue, sky blue; 5 ft. x 2 ft. or more. Magic Fountains strain shorter, to 3 ft. New Zealand hybrids resistant to powdery mildew; 5 ft. x 2 ft. or more. Pacific Giant hybrids such as 'Guinevere' double flowers lavender-pink; 5–6 ft. x 2 ft. or more.

D. grandiflorum—branched, bushy plant with loose stems; sky blue flowers. Blooms all summer. Doesn't need coddling. 24 in. x 12 in. Zones 3–9. Another parent of hybrids. 'Blue Butterfly' deep blue.

D. nudicaule—airy plant; red flowers. Blooms mid-summer. Tender; can die off in a cold winter. 24 in. x 8 in. Zones 5–8.

Dianthus

(carnation, cheddar pink, pink) *Caryophyllaceae*

Small, hardy, often fragrant; cushiony growth and needlelike evergreen foliage. Flowers often fringed, sometimes double. Easy to hybridize—sometimes does it on its own—so selection of cultivars is large and often changing.

CULTIVATION

Full sun and well-drained, normal to slightly alkaline soil (avoid overly acid soils). Will rot out if planted in poorly drained soil. Zones 4/5-9.

GARDEN USE

Although the fragrance of some is so strong you can smell it many feet away, it's best to plant where your nose can easily get to it, because you'll want to sniff it every time you walk by. Enjoys the slight alkalinity near concrete, so useful near pathways and in rockeries. Good companions include thrift (matching forms) and hardy geraniums.

CULTIVARS

'Agatha' deep pink, semidouble with dark eye. 'Bath's Pink' soft pink. 'Dad's Favourite' white, semidouble, dark eye and edged in maroon. 'Dainty Dame' white with large maroon center. 'Doris' salmon-pink, semi-double. 'Eastern Star' bright red, dark eye. 'Essex Witch' rosy, semidouble. 'Freuerhexe'/'Firewitch' raspberry-red. 'Inchmery' pale pink. 'Little Jock' semidouble, red eye; 4 in. high. 'Mrs. Sinkins' white, double. 'Pike's Pink' soft pink, double; 6 in. high. 'Sops in Wine' white infused with wine-red. 'Spotty' rose, each petal with white spot; 6 in. x 6 in. 'Tiny Rubies' deep pink, double.

Dianthus 'Bath's Pink'

Dicentra spectabilis

Dicentra

(syn. Lamprocapnos; bleeding heart) *Fumariaceae*

Woodland plants that offer good variety in texture and form. Well divided, ferny foliage. Delicate, arching stems lined with little "purses," usually pink.

CULTIVATION

Shade to part shade and well-drained, humusy soil. Even the Western native, *D. formosa*, prefers some moisture retention during the summer or it will go dormant. Many grow from tubers or rhizomes.

GARDEN USE

A herald of spring; plant with other spring flowers such as lungworts and *Epimedium*. Astilbe make good watering companions.

SPECIES

D. eximia (fringed bleeding heart)—blue-gray foliage divided many times; rose-pink flowers in terminal clusters instead of lined out. Blooms midspring to early summer. 24 in. x 18 in. Zones 4–8. 'Snowdrift' white flowers.

D. formosa—glaucous, well-divided foliage; pink flowers clustered near end of stems. Blooms mid- to late spring. Shade. 18 in. x 36 in. Zones 4–8.

D. scandens—a vine (unusual); flowers yellow elongated "hearts." Blooms summer. Part shade. 36 in. x 36 in. Zones 6–8.

D. spectabilis—same type of ferny foliage as *D. formosa*, but bigger, and on large, impressive plants; arching stems lined with flowers. Blooms late spring to early summer. 4 ft. x 1.5 ft. Zones 3–9. 'Alba' white flowers. 'Gold Heart' bright yellow foliage; pink flowers.

CULTIVARS

'Bacchanal' deep red; 15 in. x 15 in. 'Burning Heart' deep red with white accent. 'Langtrees' blue-gray foliage; white flowers; summer dormant; 12 in. x 18 in. 'Luxuriant' cherry red; 15 in. x 15 in.

Dierama pulcherrimum

Dierama

(angel's fishing rod) *Iridaceae*

Mounds of grassy foliage produce long, thin, elegantly arching stems from which dangle pink, bell-shaped flowers in summer. The sight of a slight breeze stirring them touches a gardener's heart.

CULTIVATION

Full sun and well-drained, continually moist soil. Grows from corms.

GARDEN USE

Two species in cultivation; the larger comes in varying hues. Grow pondside to see its reflection in the water or in an uncrowded bed where its form can be admired. Easy to cross, and so named cultivars, in a range of pink, red, and peach shades, change often.

SPECIES

D. *dracomontanum* (syn. **D.** *pumila*)—blooms summer. 24 in. x 24 in. Zones 8–9.
D. *pulcherrimum*—blooms summer. 3–5 ft. x 3–5 ft. Zones 8–9. Var. *album* white flowers.

Digitalis

(foxglove) *Scrophulariaceae*

Tall, with mostly basal, long, oval, pointed leaves. Spikes of funnel-shaped pendant flowers held close to the stem; often a jaunty nod at tip-tops of stems. Grows a rosette of leaves in fall, ready to send up flower stalks come spring.

CULTIVATION

Full sun to part shade and well-drained soil. Common foxglove (***D. purpurea***) can be a biennial (lasting two years) or a short-lived perennial; either way, let it go to seed and you'll have it for a long time, which isn't necessarily bad, as long as you know about it.

GARDEN USE

Good foil to usual mounds of color, especially hardy geraniums for early summer accompaniment; include *Rudbeckia* and aster 'Mönch' to fill in gaps left after foxgloves finish.

NOTES

Patrol for cutworms in spring. Common foxglove is the source of the drug digitalis.

Digitalis ferruginea

SPECIES AND CULTIVARS

***D.** ferruginea* (rusty foxglove)—evergreen leaf base; rusty copper flowers. Blooms early to midsummer. 4 ft. x 1.5 ft. Zones 4–8.

***D.** grandiflora* (yellow foxglove)—basal foliage; stems of yellow flowers with brown-speckled throats. Blooms early to midsummer. 36 in. x 18 in. Zones 3–8.

***D.** x *mertonensis* (strawberry foxglove)—basal foliage; bright pink flowers. Blooms summer. 36 in. x 12 in. Zones 3–8.

***D.** obscura*—narrow leaves; orange-brown flowers. Blooms late spring. Sun; not a woodland plant. 4 ft. x 1.5 ft. Zones 4–8.

***D.** purpurea*—basal leaves; spikes of flowers in white, pink, lavender, or purple. Blooms early summer. 3–6 ft. x 2 ft. Zones 4–8. Foxy group to 30 in. high. 'Sutton's Apricot'.

Disporum hookeri

Disporum

(syn. Prosartes; fairy bells) *Convallariaceae*

Woodland plant with dark green foliage. Small, white, pendant flowers appear in leaf axils. Yellow or orange fruit later.

CULTIVATION

Shade to part shade and well-drained soil. Takes summer-dry soils well. Zones 4–9.

GARDEN USE

Good companion to hostas, wild ginger, and epimediums. All species bloom in late spring and early summer.

SPECIES AND CULTIVARS

D. *hookeri*—oblong leaves; greenish-white flowers. 36 in. x 18 in.

D. *sessile*—leaves grow directly attached to branch, with no leaf stem. 24 in. x 24 in. 'Variegatum' white streaked edges.

D. *smithii*—large leaves. 24 in. x 12 in.

Dryopteris erythrosora

Dryopteris erythrosora
(autumn fern) *Dryopteridaceae*

Deeply cut fronds begin copper-red in spring and age to green with still a tinge of color. Ferns are nonflowering plants.

CULTIVATION
Part shade and well-drained soil. Zones 6–9.

GARDEN USE
Combines to great effect in shade with hostas or under limbed-up shrubs. 24 in. x 15 in.

Echinacea

(coneflower) *Asteraceae*

Magenta daisy flowers that sit face-up atop tall, dark stems. Many selections have a prominent brown center and ray flowers (the petals) that point down. Leaves oblong, pointed at ends, and somewhat hairy or sticky to the touch.

CULTIVATION

Full sun and well-drained soil. Not totally drought tolerant plants; may need supplemental summer water. Cut back by a third in spring to shorten flower stems (or just select a short cultivar). Deadhead for continuous flowers, or leave seedheads for finches to snack on.

GARDEN USE

Use in middle of island planting or parking strip or toward back of one-sided beds. Summer-flowering partners include garden penstemon and verbascum.

NOTES

The number of cultivars increases exponentially every year, and not all stand the test of time. Be your own judge as to their worthiness.

Echinacea purpurea

SPECIES

E. angustifolia—narrow basal leaves; purple-pink flowers with high brown center. Blooms early summer. 4 ft. x 1.5 ft. Zones 4–9.

E. pallida—similar to E. angustifolia but with fewer, strappier petals. Blooms summer. 4 ft. x 2 ft. Zones 4–8.

E. paradoxa (yellow coneflower)—smooth, lance-shaped leaves; yellow flowers with drooping petals and high brown center. Blooms late spring to early summer. 36 in. x 24 in. Zones 5–8.

E. purpurea (purple coneflower)—lance-shaped, toothed leaves; purple flowers with drooping petals and high brown center. Blooms midsummer to early fall. 36 in. x 36 in. Zones 3–9.

CULTIVARS

Pink/red: 'Doppelganger' second-story tuft of petals appears on top of center disk, but possibly not until second season. 'Magnus' flowers 7 in. across. 'Pica Bella' watermelon pink; 2 ft. 'Rubinstern' shorter petals horizontal. White: 'Green Jewel'. 'White Swan'. Yellow/orange: 'Aloha' bright yellow. 'Harvest Moon' pale gold. 'Hot Papaya' double orange-red.

Echinops

(globe thistle) *Asteraceae*

Large basal leaves deeply divided and rather spiny on tips. Sturdy stems rise above foliage, topped with balls of tiny blue flowers.

CULTIVATION

Full sun and well-drained, lean soil. Grow in a hot, sunny situation; summer dry soil is fine. Cut back after flowers finish, and in mild weather another flush of leaves will emerge. A few flower stems may show, but fall is too late for them to develop well.

GARDEN USE

Architectural plant for the dry, sunny garden. Blooms midsummer. Garden penstemon, ornamental grasses such as bronze sedge, and salvias all make good companions. Don't plant a small, delicate thing near globe thistle, as its leaves will smother it.

NOTES

Bees love globe thistle.

Echinops ritro

SPECIES AND CULTIVARS

E. bannaticus—4 ft. x 2 ft. Zones 5–9. 'Blue Globe' balls of blue 2.5 in. across. 'Taplow Blue' pale blue.
E. ritro—golfball-sized flower heads. 4 ft. x 4 ft. 'Veitch's Blue' smaller, to 36 in. high.
E. sphaerocephalus 'Arctic Glow' (white globe thistle)—white flowers; red stems to 36 in. x 36 in. Zones 3–9.

Epimedium
(barrenwort, bishop's hat) *Berberidaceae*

Usually evergreen clumping plants, some spreading gradually by underground rhizomes. Leaf stems all arise from the ground (not a central stem); compound leaves have heart-shaped leaflets, some with slightly spiny tips around edges. New growth is often bronze or red-tinted, and some color remains for a few species. Sprays of small, spurred flowers on wiry stems are held above foliage.

CULTIVATION

Shade to part shade and well-drained soil. Does not need supplemental water with a good mulch. Although most epimediums are evergreen in our climate, many gardeners tidy them up by cutting leaves back to the ground in late winter before flower stems emerge. Do not plant under a small, new tree that throws no shade; wait until the tree is big enough to create shade—otherwise you will fry the epimediums. Zones 5–9.

GARDEN USE

Invaluable for dry shade, including under shrubs and trees, for foliage as much (or sometimes even more than) for flowers. Blooms early to midspring, then followed by new foliage. Wood aster, hellebores, hostas, and London pride saxifrage all make good companions.

SPECIES AND CULTIVARS

E. grandiflorum—'Lilafee' violet; 15 in. x 15 in. 'Rose Queen' strawberry pink; 12 in. x 12 in.

E. x *perralchicum*—bronze new growth; flowers pale yellow. 16 in. x 24 in. 'Frohnleiten' flowers deeper yellow.

E. pinnatum subsp. *colchicum*—fewer spines on leaves; yellow flowers. 12 in. x 12 in.

E. pubigerum—flowers creamy white, pink inside. 18 in. x 18 in.

E. x *rubrum*—new growth tinged red, copper in fall; flowers rose with white spurs. 12 in. x 12 in.

E. x *versicolor*—spiny, toothed foliage red in spring. 12 in. x 12 in. 'Sulphureum' bright yellow. 'Versicolor' rose and pale yellow.

E. x *warleyense*—bright orange-yellow flowers. Large plant; 20 in. x 30 in. 'Oranje Koningin'/'Orange Queen' deeper orange.

E. x *youngianum*—reddish new growth; pink flowers. 12 in. x 12 in. 'Niveum' white flowers. 'Roseum' rose flowers. 'Yenomoto' white.

Epimedium x perralchicum

Erigeron karvinskianus

Erigeron
(fleabane) *Asteraceae*

Small, clumping plants with lobed leaves and daisy flowers in white or pink.

CULTIVATION
Full sun and well-drained soil. Zones 5–8.

GARDEN USE
Looks fabulous at the base of a fountain or birdbath, at the top of a wall, or growing in cracks in concrete steps. Blooms early to midsummer and longer. Charmers.

SPECIES AND CULTIVARS
E. karvinskianus (syn. 'Profusion')—flowers in pink or white on same plant. 12 in. x 36 in. 'Rosa Juwel'/'Pink Jewel' semidouble, bright pink; 24 in. x 18 in.

Erodium

(heron's bill) *Geraniaceae*

Similar to hardy geraniums with ferny foliage and small flowers.

CULTIVATION

Full sun and well-drained soil. Zones 6–8.

GARDEN USE

Small plants big on charm are best seen at the front edge of the border, in rockeries, and tumbling out of containers. Good for the dry garden. Blooms late spring.

SPECIES AND CULTIVARS

E. cheilanthefolium—green-gray foliage; pale pink flowers. 'White Pearls' white flowers.
E. chrysanthum—soft mound of silver, ferny foliage; short stems of butter-yellow, single, open flowers above foliage. Good dry garden plant. 5 in. x 12 in. 'Pickering Pink' bicolor pink flowers.

Erodium chrysanthum

Eryngium amethystinum 'Sapphire Blue'

Eryngium

(sea holly) *Apiaceae*

Plant with coarse texture, spiny leaves, and thistle-like flower heads makes an impression in the garden. Foliage often steel-colored. Flowers—which look like knobs of silver-blue—often surrounded by large, showy bracts resembling Elizabethan collars.

CULTIVATION

Full sun and well-drained soil. Summer dry soil is good. Wear protection in winter when it's time to cut it back.

GARDEN USE

Use in middle of plantings seen from all sides or at back of borders. Most bloom mid- to late summer. Most do well combined with plants of other colors, similar to globe thistle.

SPECIES AND CULTIVARS

E. agavifolium—thick, wide, swordlike leaves from a central point, leaves lined with spiny tips; flowers green-white on tall stems. 5 ft. x 2 ft. Zones 6–9.

E. alpinum—heart-shaped leaves, gray stems; flowers steel-blue. Blooms midsummer. 28 in. x 18 in. Zones 5–8.

E. amethystinum—violet-blue flowers. 28 in. x 28 in. Zones 3–8. 'Sapphire Blue' more steel blue.

E. bourgatii—slightly more refined; white veins on leaves; silver-blue flowers. 18 in. x 12 in. Zones 5–9.

E. planum (flat sea holly)—dark green, lobed leaves; steel-blue flowers. 36 in. x 18 in. Zones 5–9. 'Blue Hobbit' 12 in. x 10 in.

E. varifolium (Moroccan sea holly)—dark, marbled foliage; evergreen. 16 in. x 10 in. Zones 5–9. 'Miss Marble' foliage variegated white.

E. yuccifolium (rattlesnake master)—bristly, sword-shaped, spiny leaves. 4 ft. x 2 ft. Zones 4–8.

E. x *zabelii* 'Big Blue'—intense blue flowers. 30 in. x 20 in.

Erysimum 'Julian Orchard'

Erysimum
(wallflower) *Brassicaceae*

Evergreen mounding plants with green or gray-green foliage. Most have clusters of fragrant, four-petaled flowers.

CULTIVATION
Full sun and well-drained soil. Zones 7–9.

GARDEN USE
Blooms midspring into early summer. Combine wallflowers in the sunny border with potentilla, crocosmia, and other summer bloomers.

SPECIES
E. cheiri (syn. *Cheirianthus cheiri*)—yellow, orange, and brown; more flower colors available now in hybrids. Short-lived. 18 in. x 12 in.
E. linifolium—linear leaves, forming a shrubby mound. 28 in. x 10 in. 'Variegatum' creamy variegation to leaves.

CULTIVARS
'Apricot Twist' apricot and orange; 18 in. x 18 in. 'Bowles Mauve' blooms forever, does not set seed), and dies of exhaustion in three years; 30 in. x 24 in. 'Constant Cheer' copper-red; 24 in. x 24 in. 'Julian Orchard' rose, lavender, and purple; 18 in. x 18 in. 'Fragrant Star' yellow leaf margins; yellow flowers; 24 in. x 24 in. 'Wenlock Beauty' multicolored shades of mauve, red, and apricot; 18 in. x 18 in.

Eupatorium

(joe-pye weed) *Asteraceae*

Tall plants with dark stems. Flowers, usually pink, in tight, domed clusters at tops of stems.

CULTIVATION

Full sun and well-drained soil. Zones 4–9.

GARDEN USE

Midsummer flowers on tall plants. Surround with mounds of 'Mönch' or 'Wunder von Stafa' asters and small, spikey *Phormium* 'Jack Spratt'.

SPECIES AND CULTIVARS

E. cannabinum 'Flore Pleno'—leaves coarsely toothed and divided into five parts; double form sterile, doesn't become weedy, plus flowers more showy. 4 ft. x 2 ft.

E. maculatum—Atropurpureum group dark stems; pink domes of flowers; 8 ft. x 3 ft. 'Gateway' dark stems; mauve-pink domes of flowers; 7 ft. x 3 ft.

E. rugosum 'Chocolate'—a different take on Eupatorium; foliage flushed dark purple-brown; clusters of white flowers. Part shade. 5 ft. x 2 ft.

Eupatorium purpureum subsp. *maculatum* 'Gateway'

Euphorbia

(spurge) *Euphorbiaceae*

Distinctive collection of plants with green or blue-green foliage and eye-catching stems of flowers—although showiest parts of floral stems are chartreuse bracts that cup the tiny flowers. Mostly herbaceous. Lots of Euphorbia *hybridizing is being done—by people or the plants—so it's getting difficult to set cultivars straight according to where they belong.*

CULTIVATION

Full sun and well-drained soil. Some can be pests if given too much tender care, so keep the hose away in summer. Cut back flower stems after bloom to tidy up the plants and keep them from reseeding. Be aware of the plants that come up from running roots, so that you can put them where both you and the plant are happy.

GARDEN USE

Useful for almost every garden situation, for architecture, color, and form. Mostly spring blooming; brings a certain zing to a garden full of pink rhododendrons and azaleas. Large species should be given the room they need, so consider size and appearance of the plant and your garden.

NOTES

The milky sap in *Euphorbia* stems is caustic. Wear gloves and goggles when cutting.

SPECIES

E. amygdaloides (wood spurge)—orange-tinted foliage. Sun to part shade. 36 in. x 12 in. Zones 6–9. 'Purpurea', 'Rubra' purple-flushed foliage. Var. *robbiae* (Mrs. Robb's spurge) evergreen, glossy foliage, shrubby appearance; great for dry shade; 36 in. x 36 in.

E. characias—architectural stems like pillars; gray-green foliage. Blooms early spring. 4 ft. x 4 ft. Zones 7–10. 'Glacier Blue' leaves edged in cream. 'Lambrook Gold' flowers/bracts more yellow.

E. griffithii—blue-green leaves, dark stems. Blooms midspring. Stoloniferous. 36 in. x 24 in. Zones 4–9. 'Dixter' red bracts, stems, foliage. 'Fireglow' orange bracts and stems.

Euphorbia griffithii 'Fireglow'

E. x *martinii*—red stems, purple new growth; red eyes (the real flowers) to floral stems. 36 in. x 36 in. Zones 7–10. 'Ascot Rainbow' green, gold, and lime variegation with pink tints.

E. myrsinites (myrtle spurge)—prostrate plant with short blue leaves; stems snake around. Blooms mid-spring. 4 in. x 24 in. Zones 5–8.

E. polychroma—sulfur yellow floral stems. Blooms mid- to late spring. Takes low summer water. 16 in. x 24 in. Zones 4–9.

CULTIVARS

'Blue Haze' bushy evergreen, blue-green; 24 in. x 24 in.; zones 5–9. 'Jade Dragon' evergreen, blue-green leaves with some purple; blooms late winter; sun to part shade; 30 in. x 30 in. 'Redwing' red-tinged new growth, red stems and buds; 24 in. x 24 in.; zones 7–10.

Filipendula

(meadow sweet, queen of the prairie) *Rosaceae*

Elegant plants with divided foliage similar to astilbe or goatsbeard. Flowers a frothy plume of pink.

CULTIVATION

Full sun and well-drained soil. Most prefer full sun, especially in the maritime Northwest. Provide regular water during dry summer weeks.

GARDEN USE

Good summer bloomer, but needs to be sited with other plants that need regular water. *Lysimachia* and cardinal flowers make good companions.

SPECIES

F. purpurea—divided leaves; terminal carmine flowers in flat clusters on red stems. Blooms midsummer. Part shade. 4 ft. x 2 ft. Zones 4–9. 'Elegans' magenta pink, frothy. 'Nana' 10 in. high.

F. rubra (queen of the prairie)—divided leaves; large pink plumes of flowers. Blooms early to midsummer. Sun to part shade. 6 ft. x 4 ft. Zones 3–9. 'Venusta' purple-pink; 4 ft. x 4 ft.

F. ulmaria—divided, coarsely toothed leaves; creamy white, fragrant flowers in flat clusters. Blooms mid-summer. 6 ft. x 2 ft. Zones 3–9. 'Aurea' yellow spring foliage turns green. 'Flore Pleno' double flowers. 'Variegata' foliage irregularly splashed creamy white.

F. vulgaris (dropwort)—ferny foliage; white flowers. Blooms early summer. Requires less water than most. 24 in. x 18 in. Zones 4–9. 'Flore Pleno' double. 'Multiplex' double flowers; smaller than species, 15 in. x 18 in.

CULTIVAR

'Kahome' ferny foliage; fragrant, rose-pink flowers; blooms early summer; 18 in. x 12 in.; zones 3–8.

Filipendula rubra

Gaillardia 'Fanfare'

Gaillardia

(blanket flower) *Asteraceae*

Daisy flowers in warm shades of yellow, orange, maroon, and brown. Patterns of colors on flowers led to its common name. Leaves are gray-green and lance-shaped.

CULTIVATION

Full sun and well-drained soil. Zones 3–8.

GARDEN USE

Gaillardia speaks of summer. Looks good against a sunny wall or among red penstemon and salvias. Blooms early summer to fall.

CULTIVARS

'Amber Wheels' golden fringed petals; 36 in. x 24 in. 'Burgunder' wine red. 'Fanfare' each petal a quill, red base, yellow tip. 'Kobold'/'Goblin' deep red, yellow border; 12 in. x 18 in. 'Oranges and Lemons' orange centers with yellow tips; 30 in. x 18 in.

Gaura lindheimeri

(gaura) *Onagraceae*

Long, lax flowering stems from a central rosette of leaves.
Four-petaled, white flowers that keep coming all summer.

CULTIVATION

Sun and well-drained soil. From sunny Texas, it comes looking for what it's difficult for us to give—heat. Hardy, yet probably uncomfortable with heavy, wet winter soils, so don't be surprised if you have to replace it every couple of years. Once established, it needs no summer water. Cut flower stems back in winter without disturbing the clump of foliage. Zones 6–9.

GARDEN USE

Plant on a slope to take advantage of its sprawling nature. Often blooms all summer. 5 ft. x 3 ft.

CULTIVARS

'Corrie's Gold' gold variegation; white flowers. 'Crimson Butter-flies' dark pink. 'Passionate Rainbow' green, pink, and cream foliage; dark pink flowers. 'Pink Cloud' more upright. 'Whirling Butterflies' red stems; pink buds, white flowers; to 36 in. x 36 in.

Gaura lindheimeri 'Whirling Butterflies'

Gentiana asclepiadea

Gentiana

(gentian) *Gentianaceae*

Specialty plants of small stature and picky culture. Trumpet-shaped, incredibly blue flowers are the lure, but be prepared to wait a couple of years for plants to settle in before you see the show.

CULTIVATION

Warm but shady placement and moist, well-drained soil. Regular water.

GARDEN USE

Plant in the shady border with astilbes and Rodgersia.

SPECIES

G. *asclepiadea* (willow gentian)—elegantly upright to arching stems, long narrow leaves; deep blue, bell-shaped flowers. Blooms late summer. 36 in. x 18 in. Zones 6–9. Var. *alba* white.

Geranium

(cranesbill) *Geraniaceae*

Often called "hardy geraniums" to distinguish the genus from Pelargonium, which consists of mostly tender plants grown in summer. Obvious difference is growth habit: Geranium grows a rosette of leaves; Pelargonium grows from a stem. Five-petaled, open-faced flowers. Wide variety of plants with flowers in shades of pink, blue, purple, or white; mostly herbaceous.

CULTIVATION

Sun to shade and moist to dry soil, depending on species. Cut back once-flowering, herbaceous plants after they bloom to produce new flush of foliage and tidy things up. Cut down herbaceous plants in late winter. Zones 5–8, except as noted below.

GARDEN USE

Indispensable, easy-care genus that provides shots of color from early summer into fall and a variety of forms and textures—from perfect mounds to weavers that knit together a garden. Good with roses, ornamental grasses, and hebes in sun; hostas, sedges (*Carex*), and Jacob's ladder (*Polemonium*) in part shade. A few selections—such as *G. endressii* 'Wargrave Pink'—make good evergreen ground covers in difficult situations, such as under trees where its persistence is appreciated.

SPECIES

G. x *cantabrigiense*—mat-forming, deeply divided leaves; red fall color. Blooms early to midsummer. Sun or dry shade. 10 in. x 36 in. 'Biokovo' white with pink center. 'Cambridge' violet. 'Karmina' magenta. 'St. Ola' white.

G. cinereum—compact rosette of small, deeply cut, gray-green leaves; 1-in. pink flowers with darker netting; could get lost—keep it near a path. Blooms late spring to summer. Sun to part shade. 6 in. x 12 in. 'Ballerina' light pink. 'Laurence Flatman' rose with dark eye.

G. clarkei—mound of dissected foliage; 2-in. open flowers all summer. Sun to part shade. 20 in. x 20 in. 'Kashmir Blue' blue. 'Kashmir Pink' pink. 'Kashmir Purple' purple. 'Kashmir White' white.

G. dalmaticum—small, trailing; divided leaves less than half in. across; small flowers pink; sweet

Geranium cinereum 'Laurence Flatman'

appearance; good for rockery. Blooms late spring. 4 in. x 8 in.

G. harveyi—mat-forming or trailing; tiny silver, cut leaves; small, soft pink flowers; use as ground cover or at top of a wall. Blooms summer. 6 in. x 36 in. or more. Zones 7–9.

G. himalayense—dense mound of broad, glossy leaves up to 8 in. across, deeply divided; open, flat, 2-in.-wide flowers. Blooms late spring to summer. Sun to dry shade. 18 in. x 24 in. 'Baby Blue' pale blue with dark veining. 'Gravetye' light blue. 'Irish Blue' pale blue with purple center. 'Plenum' double; violet.

G. macrorrhizum (bigroot geranium)—evergreen; broad leaves deeply lobed, felted, apple-green with claret tones in winter; small clusters of pink or magenta flowers. Blooms late spring. Excellent in dry shade, such as under trees; spreads by thick rhizomes, but not invasive. 24 in. x 36 in. 'Album' white. 'Bevan's Variety' magenta. 'Ingwersen's Variety' light pink. 'Spessart' dark pink. 'Variegatum' foliage with irregular creamy variegation.

G. x *magnificum*—mound of medium green foliage, leaves 3 in. wide, shallowly cut; intense violet flowers. Blooms early summer. 24 in. x 24 in.

G. phaeum (mourning widow)—leaves with dark spots, sometimes splashed with white; flowers dark purple, small. Blooms late spring. Part shade. 18 in. x 12 in. 'Album' white. 'Lily Lovell' larger flowers. 'Samobor' dark spots join to form ring on leaves.

G. pratense—mounding; lavender-blue flowers; known mostly from its cultivars. Blooms all summer. 24 in. x 24 in. Midnight Reiter strain dusky, dark foliage. 'Mrs. Kendall Clark' lavender flowers with veining.

G. renardii—rounded, slightly lobed leaves with thick, slightly rough texture; light mauve veined flowers; makes a tidy mound. Blooms late spring. Sun. 24 in. x 24 in. 'Phillipe Vapelle' (hybrid) lavender. 'Whiteknights' white.

G. x *riversleaianum*—gray-green leaves; flowers pink with darker veins. Blooms for extended period into late summer. Sun. 12 in. x 36 in. 'Mavis Simpson' shell pink, dark veins. 'Russell Pritchard' magenta.

G. sanguineum—small, well-cut leaves; small red flowers; dainty look. Blooms all summer. 10 in. x 10 in. 'Album' white. 'Alpenglow' red. 'Cedric Morris' carmine. 'Glenluce' pink. 'Max Frei' red. 'New Hampshire Purple' purple.

G. wallichianum 'Buxton's Variety' ('Buxton's Blue')—marbled foliage; large blue flowers with white center. Blooms late summer. Sun to part shade. 12 in. x 36 in.

CULTIVARS

'Ann Folkard' lax weaver; chartreuse foliage; 1-in. bright magenta flowers with dark eye; blooms all summer; full sun or part shade; 24 in. x 36 in. 'Brookside' deeply cut leaves; large blue flowers with white eye; sterile; blooms summer; 18 in. x 18 in. 'Confetti' spreading mound of foliage splashed with white; small pink flowers; blooms summer; 8 in. x 30 in. 'Johnson's Blue' large lobed leaves; big blue flowers with red veining; blooms summer; 18 in. x 24 in. 'Nimbus' loose mound of dissected foliage; flowers lavender with dark veining; blooms all summer; 24 in. x 36 in. 'Orion' big blue saucer flowers with red veins; blooms summer; 18 in. x 18 in. 'Patricia' large magenta flowers; blooms all summer; 36 in. x 36 in. 'Rozanne' big blue flowers with wide white eye; blooms many months; sterile; 18 in. x 18 in. 'Salome' dark veined pink flowers with deep violet center; blooms summer; 18 in. x 24 in. 'Spinners' blue flowers; blooms much of summer; don't let it smother neighbors; 36 in. x 36 in. 'Stanhoe' dark cast to foliage; small, light pink flowers; blooms over long period in summer; although large, stays compact; 18 in. x 36 in.

Geum 'Mango Lassi'

Geum

(avens, old man's whiskers, prairie smoke) *Rosaceae*

Single or double, small, roselike flowers on tall stems; usually only once-blooming. Basal foliage—long, lobed, slightly hairy leaves growing from a central rosette—remains as a placeholder in the garden.

CULTIVATION

Full sun (except as noted) and well-drained soil. Cutting back stems of spent flowers may encourage reblooming. Divide in fall by digging out and separating new rosettes from main plant.

GARDEN USE

Cottage garden plant. Most bloom in mid- to late spring, so combining them with later-flowering plants, such as veronicas, helps extend the show in a particular bed. Keep small plants near paths or at tops of walls; larger plants can go deeper in beds, where long flower stems may bob and weave among other plants.

SPECIES AND CULTIVARS

G. coccineum 'Eos'—orange flowers against golden foliage; blooms late spring. 14 in. x 18 in. Zones 5–8.
G. rivale (water avens)—glossy leaves more round, tidy rosettes; flowers nod and do not fully open; flower covering is red, which contrasts with apricot flowers. Blooms late spring to midsummer. Part shade. 8 in. x 24 in. Zones 3–8. 'Coppertone' tawny apricot. 'Leonard's Variety' copper.
G. triflorum (old man's whiskers, prairie smoke)—small plants used to hardships of high altitudes; flowers nod, then turn upright as whispy seedheads develop. Blooms summer. 10 in. x 12 in. Zones 1–8.

CULTIVARS

Flowers 1.5 in. across; all generally 24 in. x 24 in. 'Beech House Apricot' apricot. 'Blazing Sunset' orange flowers 2 in. wide throughout summer. 'Mango Lassi' shades of apricot, pink, orange. 'Red Wings' scarlet. 'Starkers Magnificum' double orange. 'Totally Tangerine' single orange throughout summer.

Gypsophila paniculata 'Pink Fairy'

Gypsophila

(baby's breath) *Caryophyllaceae*

Airy growth. Small leaves and wiry stems. Small pink or white flowers.

CULTIVATION

Full sun and well-drained soil.

GARDEN USE

Known mostly as an ingredient in floral arrangements; comes as a surprise to many gardeners that there are selections good for the garden. Good contrast to other plants' large flowers and heavy foliage.

SPECIES AND CULTIVARS

G. paniculata—basal foliage stays in background; stems heavy with clusters of small white flowers rise up. Blooms midsummer. Stake or grow with supporting neighbors. 4 ft. x 4 ft. Zones 4–9. 'Bristol Fairy' double white. 'Pink Fairy' double pink flowers; 18 in.

G. repens—creeping; looks best growing in or at top of a wall where it can spill over; pink flowers. Blooms all summer. 8 in. x 20 in. Zones 4–8. 'Alba' white. 'Rosea' deep pink.

Hakonechloa macra

(Japanese forest grass) *Poaceae*

Deciduous, low-growing grass with a soft, shaggy texture. Most noted for its cultivars with colorful foliage that turns taupe in winter.

Hakonechloa macra 'Beni Kaze'

CULTIVATION

Part shade in well-drained, evenly moist soil. Shear back at end of winter. Zones 6–9.

GARDEN USE

Create your own *Hakonechloa* waterfall by planting several of the same selection down a slope. Combine with other shade-lovers such as hostas and Polystichum. 12 in. x 18 in.

CULTIVARS

'Albovariegata' creamy white stripes along leaves. 'All Gold' intense chartreuse. 'Aureola' leaves heavily striped with bright yellow. 'Beni Kaze' leaves turn reddish in autumn.

Helenium 'Coppelia'

Helenium autumnale
(sneezeweed; Helen's flower) *Asteraceae*

Clump-forming plants with lance-shaped leaves. Daisy flowers have wide petals that are pinked at tips and a prominent brown cone in the middle (where, as with all daisy flowers, the true flowers reside). The common name of sneezeweed is misleading; it doesn't make you sneeze.

CULTIVATION
Full sun, well-drained soil, regular summer water. Zones 4–8.

GARDEN USE
Sunny flowers bloom midsummer to fall. Grow with *Bergenia* and ornamental grasses. Deadhead for longer bloom time.

CULTIVARS
Most selections are hybrids that grow 36 in. x 24 in. 'Butterpat' yellow. 'Coppelia' copper-red. 'Moerheim Beauty' deep red; begins blooming early. 'Sahin's Early Flowerer' burnt orange with yellow edges.

Helianthus

(sunflower) *Asteraceae*

Perennial sunflowers are news to many, although these selections have been around awhile. Tall. Leaves have a rough feel. Flowers are typical of sunflowers, although smaller than annual Helianthus.

CULTIVATION

Full sun and well-drained soil. Not drought tolerant; needs water during dry weeks of summer. Shows off what a little warmth and sun can do. Zones 6–9.

GARDEN USE

Blooms mid- to late summer. Position at back of the border or in middle of an island bed.

SPECIES AND CULTIVARS

***H.** angustifolius* (swamp sunflower)—narrow, lance-shaped leaves and hairy stems; clusters of 3-in. yellow flowers with brown center. 5–7 ft. x 4 ft. 'Gold Lace' more floriferous, shorter stature.
***H.** salicifolius* (willowleaf sunflower)—similar to H. angustifolius, but will take some dryness. 8 ft. x 3 ft. 'First Light' to only 36 in. high. 'Lemon Queen' light yellow; 5 ft. x 3 ft. 'Table Mountain' 18 in. x 18 in.

Helianthus 'Lemon Queen'

Helictotrichon sempervirens

(blue oat grass) *Poaceae*

Evergreen grass with steel blue foliage, creating fountain effect. Flowers in early summer with tan stalks held above foliage.

CULTIVATION

Full sun and well-drained soil. No supplemental water once established. Do not cut down; instead "comb" out dead foliage with a small leaf rake in late winter. Reseeds modestly. Zones 4–9.

GARDEN USE

Use with hebes, sea holly, and small conifers. 36 in. x 36 in.

Heliopsis (false sunflower) Asteraceae
Yellow daisy flowers. Leaves are lance-shaped and have toothed edges.

CULTIVATION

Grow in full sun and well-drained, not overly fertile, soil. Will also take some summer dryness. Zones 4–9.

GARDEN USE

Sunflower relative that blooms beginning in mid-summer and over a long period. Good late color.

SPECIES AND CULTIVARS

H. helianthoides—'Loraine Sunshine' leaves almost white with green veins; 30 in. x 16 in. Var. *scabra* 'Summer Sun' double flowers with light brown disk; to 36 in. x 36 in.

Helictotrichon sempervirens

Helleborus

(Christmas rose, hellebore, Lenten rose)
Ranunculaceae

Mostly evergreen and shrubby. Leaves usually divided; some have slightly spiny edges. Flowers range from single, nodding "roses" in white or purple to greenish-white clusters on strong stems. Breeding is ongoing, for gardeners have fallen in love with hellebores. Now there are flowers double purple, single apricot, and yellow, as well as flowers that look like pompons.

CULTIVATION

Part shade (except as noted) and well-drained soil. Prefers slightly alkaline soil; many gardeners now apply a mulch of limestone chips (bought from a quarry or stonemason). Although it is not drought tolerant, a good mulch helps keep watering at a minimum in the summer. Old leaves of *H. x hybridus* plants often cut off entirely before flower stems emerge in winter for a tidy look.

GARDEN USE

Provides colors and texture in the garden, especially fall through early spring. Use in containers, in the woodland garden, at edges of paths (especially concrete paths, to help with alkalinity), and as landscape features—pick your need and there's a hellebore for you (as long as you meet the plant's cultural needs).

SPECIES

H. argutifolius (syn. *H. corsicus*; Corsican hellebore)—shrubby plant with thick, light-colored stems and divided leaves with soft, spiny edges; green-white flowers emerge like a fist. Blooms late winter to early spring. 36 in. x 36 in. Zones 6–9. Janet Starnes strain white speckled foliage.

H. foetidus (stinking hellebore)—dark green leaves divided into narrow leaflets, like palm of your hand, often with a red spot at center point (foliage is fabulous feature); clusters of pale green, cup-shaped flowers. Blooms midwinter to spring. 32 in. x 18 in. Zones 6–9. Wester Flisk group deeper and more red to leaves and stems.

H. x *hybridus* (Lenten rose)—dark leaves divided into threes; nodding flowers in greenish-white. Blooms winter to early spring. 18 in. x 18 in. Zones 6–9. Formerly known as Orientalis group and *H*. *orientalis*. Myriad strains and groups, ranging from pure white to dusky purple, deep black-purple, slate blue, and even reddish purple; doubles, too; seed-grown and many without established cultivar names; buy them in bloom to select desired color.

H. *niger* (Christmas rose)—dark green, divided leaves; pure white flowers emerge just above foliage. Blooms midwinter. 12 in. x 1 in. Zones 4–8.

H. x *sternii*—mound of dark, blue-green foliage; white flowers often blushed with pink. Blooms late winter to early spring. 14 in. x 12 in. Zones 6–9.

CULTIVARS

From crosses involving *H*. *niger*, breeding hardy, winter-blooming plants with outward-facing, long-lasting flowers. White: 'Ivory Prince', 'Jacob', 'Josef Lemper'; pink: 'Cinnamon Snow', 'Pink Frost'.

Helleborus x hybridus

Hemerocallis

(daylily) *Liliaceae*

Grassy foliage, evergreen for many species. Lily-shaped flowers, each lasting only a day; some have ruffled edges, others are split into petals like a lily; many have more than one color. There are enough daylily cultivars—more than 30,000—to populate a small town, and these are only the tip of the iceberg. If the daylily bug bites you, you'll be able to find many more selections at nurseries and through mail-order.

CULTIVATION

Full sun and well-drained soil. Established plants need no supplemental summer water. After a hard winter, evergreen foliage can look a little tired; cut it all off to ground level in late winter to make way for new leaves. Divide in fall. Zones 3–9.

GARDEN USE

Hugely popular; good garden component. Besides its range of flowers, different sizes fit into any garden. Small-growing daylilies can grow along a path or accent the base of a rose. Better yet, grow a taller daylily at the base of a rose to hide its bare stems. Choose continuous bloomers for repeat flowering.

NOTES

Yes, daylily flowers are edible.

SPECIES

H. flava—fragrant, lemon-yellow flowers. Blooms spring. 24 in. x 24 in. 'Tetrina's Daughter' fragrant, yellow; ever-blooming.

CULTIVARS (BY SIZE)

Smaller: 'Happy Returns' yellow; repeats bloom; 18 in. 'Little Grapette' purple; early bloom; 18 in. 'Stella d'Oro yellow; continuous bloom; 18 in. Larger: 'Chicago Apache' red; 30 in. x 24 in. Voted best by Pacific Northwest daylily gardeners: 'Canadian Border Patrol' cream with purple edge and purple throat; blooms early to midsummer, repeats bloom. 'Custard Candy' creamy yellow with maroon throat; repeats bloom. 'Janice Brown' light pink with rose eye; blooms all summer. 'Joan Senior' white 6-in. flowers; repeats bloom.

Hesperantha—see *Schizostylis*

Hemerocallis 'Chicago Apache'

Hesperis matronalis

Hesperis matronalis
(dame's rocket) *Brassicaceae*

Tall stems with oblong to lance-shaped leaves and domes of fragrant lavender flowers that, as members of the cabbage family, have four petals each.

CULTIVATION

Full sun to part shade and well-drained soil. After flowering, plants go to seed and foliage gets powdery mildew. Cut back hard or pull out and let new plants take over. Zones 4–9.

GARDEN USE

Blooms mid- to late spring. Old-fashioned flower often labeled "passalong plant," meaning it either reseeds or increases easily. Always prompts questions from nongardeners or new gardeners eager to plant something that's pretty and smells good. It will grow where it wants to grow in your garden, if you let it reseed. Then, you can edit out the unwanted plants. 36 in. x 18 in.

Heuchera

(alumroot, coral bells) *Saxifragaceae*

Although old selections are still worthy plants—rounded, evergreen leaves in a low rosette, with thin stems topped with airy clusters of bright red flowers—these days it's all about foliage. The number of cultivars is legion and like the multiplying broom in the sorcerer's apprentice, new names are added each year.

CULTIVATION

Full sun and well-drained soil; *Heuchera* grown for foliar effect will take part shade. Once established, does not need supplemental water in summer. A few small, dirty white flowers above purple leaves are nothing to write home about, so feel free to cut off flower stems in favor of foliage. Cut back old foliage in late winter. As plant increases by offsets, every few years the center of the plant gets woody and bare; dig up and replant young offsets and toss the rest on the compost pile. Zones 4–9.

GARDEN USE

Long the denizens of rockeries around the region; red flowers above green leaves a pretty sight. Use dark *Heucheras* in part shade because in deep shade, it is harder to see its deep purple leaves. Use brightly variegated grasses nearby, such as Japanese forest grass (*Hakonechloa macra* 'Aureola') or *Dicentra* 'Gold Heart'.

NOTES

Root weevils can be a problem; use mechanical controls by going hunting at night.

SPECIES

H. americana (alumroot)—new foliage with purple veins. Blooms early summer. Sun. 18 in. x 12 in.
H. sanguinea—heart-shaped basal leaves; bright red flowers on slender stems. Blooms late spring. Sun. 18 in. x 12 in. 'Snow Storm' leaves stippled with white. 'Splendens' larger flowers.

CULTIVARS

18 in. x 24 in.; zones 4–8. 'Green Spice' green leaves with dark veins and overall pewter cast. Red: 'Amber

Heuchera 'Green Spice'

Waves' foliage flushed apricot, yellow, and gold. 'Marmalade' apricot, gold, and deep rose. 'Peach Flambe' red with peach overtones. For dark foliage: 'Amethyst Myst' plum foliage with silver cast. 'Autumn Haze' purple foliage. 'Crimson Curls' curly purple foliage. 'Pewter Moon' purple foliage with darker undersides. 'Pewter Veil' flushed purple foliage with metallic silver. 'Plum Pudding' purple with darker veins. 'Velvet Night' deep purple with silver veins.

x *Heucherella* an intergeneric hybrid in all ways, as it comes with *Heuchera* leaves but with the more showy flowers of *Tiarella*. 'Solar Eclipse' red-brown leaves with lime edge. 'Sweet Tea' orange leaves with dark veins.

Hosta

(funkia, plantain lily) *Liliacaeae*

Quintessential herbaceous perennial for shade. The role it fills in the garden almost excuses the number of cultivars coming out all the time, which all have slight variations in color or variegation. Growing from a rosette of leaves that begin upright and unfurl into a horizontal position. In summer, stalks of white or lavender flowers emerge from plant's center. Flowers, although attractive, are not the plant's best feature (except in those with fragrant flowers).

CULTIVATION

Shade to part shade and well-drained soil. Established plants take low summer water, especially when mulched. Divide in fall or early spring by digging up whole plant and use pruning saw to cut through crown, keeping roots with each piece. Divide only if you want to share; otherwise, let plants bulk up to an impressive state. Zones 3–9.

GARDEN USE

Given room to spread, looks fabulous under a tree or combined in the shade garden with epimediums and hellebores. Line a path with small ones; plant in sheltered beds at the front door; grow in pots.

NOTES

Slugs can turn hostas into swiss cheese overnight; use nontoxic slug control, such as beer traps or phosphorus pellets. Hostas with thick, blue-green leaves show more slug resistance.

SPECIES

H. plantaginea—shiny, deep green leaves; highly fragrant, white flowers. 20 in. x 24 in.
H. sieboldiana var. *elegans*—broad, blue-green, seersucker leaves; white flowers. 4 ft. x 4 ft.

CULTIVARS (SIZE NOTED AS SMALL, MEDIUM, OR LARGE)

Green: 'Honeybells' fragrant white flowers; medium. 'Royal Standard' glossy, deep green foliage; fragrant white flowers; medium. Variegated: 'Albomarginata' white margins; medium. 'Aureomarginata'

Hosta 'Gold Standard'

yellow margins; medium. 'Francee' white margins; medium. 'Frances Williams' blue-green leaves, yellow margins; white flowers; medium. 'Great Expectations' centers full of creamy yellow; large. 'June' gold centers, blue-green leaves; medium. 'Minuteman' pure white margins; medium. 'Patriot' white margins on narrow leaves; medium. 'Wide Brim' yellow margin; large. Yellow foliage: 'August Moon' medium. 'Gold Standard' large. 'Golden Tiara' small. 'Sum and Substance' overall chartreuse; large. Blue foliage: 'Big Daddy' blue-green; large. 'Blue Angel' large. 'Blue Cadet' small. 'Guacamole' large. 'Halcyon' large.

Impatiens omeiana
(impatiens) *Balsaminaceae*

Whorls of long, slender, dark green leaves with creamy white veins and red stems. Small apricot flowers definitely not the most important feature.

CULTIVATION

Shade to part shade and well-drained soil. Provide moderate moisture or at least a good mulch. Spreads moderately by stolons, but if it makes you nervous, plant it in a bed surrounded by concrete. Zones 6–9.

GARDEN USE

Blooms late summer. Fabulous foliar effects for the woodland garden; a standout in the shade garden. Combine with foamflowers (*Tiarella*), *Epimedium*, and wild ginger (*Asarum*). To 12 in. high and spreads.

Iris, Pacific Coast hybrid

Iris

(iris) *Iridaceae*

As old as the hills; long-lasting perennials that often evoke memories of grandmother's garden. Genus encompasses a variety of plants that generally have flower structure in common: petals divided into falls and standards. Many have broad, swordlike leaves; others have grassy foliage. May be evergreen or deciduous.

CULTIVATION

Full sun and well-drained soil, except Japanese iris, which need copious amounts of water (or to be at stream's edge). Bearded iris and Pacific Coast iris can take dry summer soils. Plant rhizomes of bearded iris at soil level, and keep competing grass and other plants away. For more specific culture, see Species and Cultivars, below.

GARDEN USE

Late-spring and summer bloomer that fades into background after planting, although foliage stays good enough to offer structure. Wide color range. Smaller iris can be grown in pots. With such a range, there are iris for any garden style under almost any conditions.

SPECIES

I. foetidissima (stinking iris, gladwyn iris)—evergreen; swordlike leaves; dull, pale violet flowers develop into seed pods that open to reveal bright red fruit that remains into winter. Blooms late spring. Sun to part shade. 24 in. x 24 in. Zones 7–9.

I. pallida (Dalmatian iris)—evergreen; grayish, swordlike leaves; branched stems with fragrant lilac-blue flowers; cultivars more common than species. Blooms late spring. 36 in. x 24 in. Zones 6–9. 'Argentea Variegata' white variegation. 'Variegata' ('Aurea Variegata') yellow variegation.

I. sibirica x *I. sanguinea* (Siberian iris)—these are crosses and recrosses from the parent species. Foliage thin, more narrow than that of bearded iris; deciduous; flowers violet-blue. Blooms late spring. Regular water; divide in fall; cut back old foliage in late winter. 4 ft. x 3 ft. Zones 4–7. 'Bennerup Blue' deep lavender-blue. 'Caesar's Brother' blue.

Iris unguicularis (Algerian iris)—evergreen foliage from rhizomes; winter-blooming with fragrant violet flowers marked yellow. 18 in. x 18 in. Zones 8–9.

CULTIVARS BY TYPE

Bearded iris—huge collections of plants divided into groups and subgroups. Blooms early summer. Divide in fall. Zones 4–9. Good garden choices in some of the different categories follow. Tall: 'Best Bet' two-toned blue. 'Dusky Challenger' deep purple-blue. 'Silverado' full, light lavender-blue. Intermediate tall: 'Maui Moonlight' ruffled lemon yellow. Dwarf: 'Serenity Prayer' cream flushed yellow with blue beard. Border: 'Batik' purple streaked with white. 'Cranapple' cranberry red. Miniature (table iris): 'Bumblebee Deelite' yellow standards, dark maroon falls.

Japanese iris—leaves stiff and upright; flat-topped look to flowers. Blooms early summer. Water-loving; divide in fall. 36 in. x 18 in. Zones 5–9. 'Fortune' white and purple. 'Loyalty' double; purple-splashed.

Pacific Coast iris—hybridized from among eleven West Coast species; often found unnamed (no cultivar) at the nursery—buy them anyway. Evergreen foliage forms an ever-widening circle; flowers in a range of lavender, purple, gold, and white with distinct markings on petals. Blooms late spring. Sun to part shade; no supplemental water in summer; in early fall when active growth begins, divide by taking new growth off from around edges. 24 in. x 24 in. Zones 8–10.

Kirengeshoma palmata

(yellow waxbells) *Hydrangeaceae*

A lanky plant with large lobed leaves and yellow funnel-shaped, drooping flowers in mid- to late summer.

Kirengeshoma palmata

CULTIVATION

Shade to part shade in well-drained soil. Zones 5–8.

GARDEN USE

Subtle but useful for the shade garden at a time when little else is blooming. Mixes well with hydrangeas and epimediums. 4 ft. x 3 ft.

Knautia macedonica 'Melton Pastels'

Knautia macedonica

(knautia) *Dipsacaceae*

Rosette of foliage—often evergreen—with long stems. Deep maroon, 2-in.-wide, pincushion flowers.

CULTIVATION

Sun and well-drained soil. Has a tendency to reseed that exasperates many gardeners; on the plus side, new plants are easy to dig out. Zones 5–9.

GARDEN USE

Blooms mid- to late summer over a long period. A good weaving plant around roses, lady's mantle, and *Euphorbia griffithii.* 32 in. x 18 in.

CULTIVARS

'Mars Midget' 16 in. x 20 in. 'Melton Pastels' mix of pink, rose, crimson. 'Thunder and Lightning' leaves edged in white.

Kniphofia

(poker plant, red hot poker, torch lily) *Liliaceae*

Base of grassy or swordlike leaves. Torches of small, tubular flowers; old-fashioned red-hot poker with two-toned orange-and-red flower now joined by selections in bright colors often in graded tones from bottom to top.

CULTIVATION

Full sun and well-drained soil; soggy soil in winter can do them in. Cut off faded flower stems to encourage more blooms.

GARDEN USE

Architectural plant with a form that foils many shorter, mounding perennials. Impressive flower stems draw attention not only of onlookers but hummingbirds too. New colors easier to combine in the garden. In colder regions, grow them in containers that can be taken in during winter.

SPECIES

K. caulescens—grassy evergreen foliage; coral and yellow flower heads. Blooms late summer. 4 ft. x 2 ft. Zones 6–9.

Kniphofia 'Bressingham Comet'

***K.** northiae*—wide blades; evergreen; coral and cream flowers. Blooms early to late summer. 5 ft. x 3 ft. Zones 6–9.

***K.** uvaria*—evergreen; coral fading to cream down the flower. Blooms summer. 4 ft. x 2 ft. Zones 6–9.

CULTIVARS (BLOOM SUMMER)

'Alcazar' coral to terra-cotta; 3.5 ft. x 1.5 ft. 'Bressingham Comet' grassy foliage; flowers fiery red fading down to yellow; 24 in. x 18 in. 'Little Maid' grassy foliage; creamy yellow flowers; 24 in. x 24 in. 'Nancy's Red' coral-red; grassy foliage. 'Primrose Beauty' lemon yellow flowers; grassy leaves; 36 in. x 24 in. 'Shining Sceptre' flowers copper fading to golden; dark stems; 36 in. x 18 in.

Lamprocapnos—see *Dicentra*

Leucanthemum x superbum 'Becky'

Leucanthemum x superbum 'Becky'

(Shasta daisy) *Asteraceae*

This is where Shasta daisies landed when the genus Chrysanthemum was broken up. Evergreen rosette of dark green, lance-shaped leaves form a sturdy stand. Of the many selections available, 'Becky' is a fine one: 3-in.-wide white flowers with yellow centers.

CULTIVATION

Sun and well-drained soil. Deadhead mercilessly. Zones 5–9.

GARDEN USE

Blooms early summer. Stands fairly upright. Grow with hardy geraniums, daylilies, and phlox. 36 in. x 36 in.

Lewisia cotyledon

native

Lewisia

(bitterroot) *Portulacaceae*

Rosettes of succulent, spoon-shaped, evergreen leaves. Bright flowers on stems held above foliage. Western native.

CULTIVATION

Full sun and sharp-draining soil. Needs to dry out after spring flowering. Zones 6–8.

GARDEN USE

Small plant best grown in terra-cotta pots or troughs. Good plant for shelves or raised areas where you can appreciate it better. Blooms late spring to early summer.

SPECIES AND CULTIVARS

L. cotyledon—often found in nurseries as strains, groups, and hybrids. Sunset strain shades of peach, orange, pink, yellow. 12 in. x 10 in. 'Little Plum' pink with hint of orange. 'Regenbogen' pink and salmon.

Liatris spicata

(gayfeather) *Asteraceae*

Long, narrow basal leaves set the stage for flower stems covered in tiny magenta flowers. Blooms from the top down.

CULTIVATION

Sun and well-drained soil. Supplemental water in summer. Zones 4–9.

GARDEN USE

Often used in florist bouquets. Blooms late summer. Unfortunately, flower stems begin to appear brown and dead before whole thing finishes blooming; place within the border with lots of other things going on. 36 in. x 18 in.

CULTIVARS

'Floristan Violett' violet. 'Kobold' mauve-pink; 24 in. x 18 in.

Liatris spicata 'Kobold'

Ligularia

(ligularia) *Asteraceae*

Dramatic, long-lasting plants if cultural needs are met. Tall, elegant flower stems rise from large basal leaves. Flowers usually yellow, sometimes shaped like daisies. What was L. 'Cristata' has been moved to Farfugium japonicum 'Crispatum' (crested, round leaves with a pink tint; odd yellow flowers; 24 in. x 24 in).

CULTIVATION

Part shade and moist soil; mulch well. Loves to be streamside, where it can get all the water it wants; even slightly boggy soil is good. Zones 5–9.

GARDEN USE

Combine with other water-lovers, such as astilbe and rodgersia.

NOTES

Slugs love *Ligularia*, so be sure to use nontoxic (beer traps, phosphorus pellets) control methods.

SPECIES

L. dentata—toothed, oval leaves; daisy flowers up to 5 in. across. Blooms summer. 4 ft. x 3 ft. Zones 5–8. 'Britt-Marie Crawford' purple-black foliage; yellow daisy flowers. 'Desdemona' and 'Othello' foliage starts out purple, turns green except for undersides. *L. przewalskii* 'Dragon's Breath'—triangular, toothed leaves; small yellow flowers on spikes. Blooms mid- to late summer. 30 in. x 24 in.

CULTIVARS

'Gregynog Gold' bright orange flowers on spikes; 5 ft. x 3 ft. 'The Rocket' golden yellow flowers; 4 ft. x 3 ft.

Ligularia dentata

Linaria purpurea 'Canon Went'

Linaria purpurea

(purple toadflax) *Scrophulariaceae*

Narrow stems coated with thin, short, gray-green leaves. Spikes of tiny snapdragon-like lavender flowers.

CULTIVATION

Sun and well-drained soil. Easy-care plant. Doesn't run, but it does reseed. On the good side, this gives you almost continuous flowering throughout summer from young, middle-aged, and settled plants. On the bad side, they tend to come up everywhere. Zones 5–8.

GARDEN USE

Blooms early summer to fall. Good cut flowers. Provides a pointy form in the garden and looks fabulous with hardy geraniums, roses, ornamental onions, and contrasting colors of crocosmia and *Rudbeckia*. 36 in. x 12 in.

CULTIVARS

'Canon Went' pink.

Linum

(flax) *Linaceae*

Mostly considered a wildflower. Foliage is variable according to species. Flowers open, blue.

CULTIVATION

Full sun and well-drained soil; short-lived in our climate that produces soggy winter soils. Zones 5–8.

GARDEN USE

A lovely, airy look to it; provides beautiful blue hues to offset all the pinks and purples we love to grow. Plant with geraniums, soapwort, and *Geum*.

SPECIES AND CULTIVARS

L. narbonense—narrow, stiff leaves; 2-in.-wide blue flowers with white eye. Blooms early to midsummer. 24 in. x 18 in. 'Heavenly Blue' dark blue; to 12 in.
L. perenne—blue-green, grassy foliage; delicate 1-in. sky-blue flowers. Blooms over long period late spring to midsummer. 24 in. x 12 in. 'Album' white. 'Blau Sapir'/'Blue Sapphire' more intense blue; to 12 in.

Linum perenne 'Blue Sapphire'

Lobelia tupa

Lobelia

(cardinal flower) *Campanulaceae*

There's more to this genus than just the well-known annual basket-stuffers; perennial Lobelia *is a tall, colorful plant for the border. Tubular, lipped flowers on sturdy stems from basal rosette of leaves. More selections being bred for tinted foliage as well as bright flower colors.*

CULTIVATION

Full sun to part shade and well-drained but continually moist soil. Some newer cultivars do well in less-than-wet situations.

GARDEN USE

Grow with *Tradescantia* or, in part shade, *Rodgersia*. Container culture is fine and helps you monitor the water situation better. *L. tupa* suits the dry garden.

SPECIES

L. cardinalis (cardinal flower)—reddish-purple stems, rosette of leaves; deep red flowers. Blooms summer. 36 in. x 12 in. Zones 3–9.

L. laxiflora (torch lobelia)—red stems, narrow leaves; narrowly tubular red flowers split to show golden throat. Blooms throughout summer. Full sun. 24 in. x 36 in. Zones 7–10.

L. tupa—broadly oblong leaves with pointed tips; spikes of long, scarlet, narrowly tubular flowers; a stunning show. Blooms late summer. Full sun. Dry summer soil OK. 6 ft. x 4 ft. Zones 8–9.

CULTIVARS (BLOOM ALL LATE SUMMER)

'Dark Crusader' dark leaves; magenta flowers; 30 in. x 18 in. 'Grape Knee-Hi' purple flowers; 22 in. x 18 in. 'Queen Victoria' purple leaves; red flowers; 36 in. x 12 in. 'Royal Fuchsia' fuchsia-pink; 36 in. x 12 in. 'Ruby Slippers' red-pink; 36 in. x 12 in.

Lupinus, Russell hybrid

Lupinus
(lupine) *Fabaceae*

Shrubby perennials that grow from a large base of palmately divided leaves. Pillars of pealike flowers in shades of pink, purple, blue, yellow, orange, and red.

CULTIVATION

Full sun and well-drained soil. Provide good air circulation to keep powdery mildew at bay. Zones 4–8.

GARDEN USE

Grow with *Rudbeckia* and late asters, which will provide a sequence of flowers. Blooms late spring and early summer. 36 in. x 16 in.

SPECIES AND CULTIVARS

Most-often grown are strains of Russell hybrids, but those have been ill bred and are regarded as having a diminished floral show and being too susceptible to powdery mildew. Breeders are going back to the original Russell strains and have added a New Generation strain that blooms for a longer period. Gallery series in blue, pink, red, and white on dwarf plants. 24 in. x 16 in.

Lychnis

(rose campion, Maltese cross, ragged robin)
Caryophyllaceae

A variety of colors and forms are contained in this one genus. Flowers generally have five petals and appear singly or in clusters at the ends of stems. Rose campion is one of those plants that appears in almost everyone's garden, just because of its tendency to lightly reseed.

CULTIVATION

Full sun and well-drained soil. **L.** *coronaria* puts up with dry summer soil. Pinch back **L.** x *arkwrightii* and **L.** *chalcedonica* early in season to encourage branching. Cut back once-flowering plants after they bloom.

GARDEN USE

Grow in the mixed border, shorter selections toward the front.

Lychnis coronaria 'Alba'

SPECIES AND CULTIVARS

L. x *arkwrightii*—oblong, pointed leaves; small clusters of scarlet-orange flowers with notched petals. Short-lived. Blooms early to midsummer. 18 in. x 12 in. 'Vesuvius' dark bronze foliage; to 12 in.

L. coronaria (rose campion)—felted gray leaves, basal rosette during winter; nearly leafless, branched stems of magenta flowers. Blooms early to midsummer. 32 in. x 18 in. Zones 4–8. 'Alba' white. Atrosanguinea group has deeper magenta shades.

L. chalcedonica (Maltese cross)—brilliant scarlet-orange clusters of small flowers with notched petals (resembling Maltese cross). Blooms early to midsummer. Stake or provide supporting companions. 36 in. x 12 in. Zones 4–8.

Lysimachia
(creeping jenny, loosestrife) *Primulaceae*

A collection of mostly moisture-loving plants of varying sizes with small, cup-shaped, often yellow flowers. Leaves variable according to species.

CULTIVATION

Sun to part shade and well-drained soil. Provide adequate water for those that need it, except where you'd prefer to limit their aggressive tendencies; cutting back on water will usually limit their spread. Cut back taller plants after bloom. Dig out sections of the ground cover or sink a barrier a few inches into the ground. Zones 4–9.

GARDEN USE

A variety of forms—tall and narrow to mat-forming—and bright colors make it easy to find a place in the garden.

SPECIES AND CULTIVARS

L. atropurpurea—lance-shaped, silver-green leaves; densely clothed spikes of purple flowers. Blooms late

Lysimachia clethroides

spring to fall. 24 in. x 24 in. 'Beaujolais' (wine loose-strife) narrow spike of bright wine color; 36 in. x 24 in.
L. ciliata—egg-shaped, 6-in. leaves; red stems fade to green; clusters of pale yellow flowers. Blooms midsummer; cut back in July to produce new growth and new red. Sun. 4 ft. x 2 ft. 'Firecracker' foliage flushed deep purple.
L. clethroides (gooseneck loosestrife)—dense spikes of tiny white flowers bend and curve to look like its common name. Blooms mid- to late summer. Don't water too much or it will be invasive. 36 in. x 24 in.
L. nummularia (creeping jenny)—ground cover that roots as it goes along; small, yellow, upturned flowers. Blooms summer. Chop it to stop it from spreading. 2 in. x indef. 'Aurea' screaming yellow foliage.
L. punctata—yellow flowers mostly in leaf axils. Blooms midsummer. A potential thug. 36 in. x 24 in. 'Alexander' variegated foliage.

Meconopsis

(poppy) *Papaveraceae*

Delicate-looking plant with single flowers in nodding blue or outward-facing yellow or orange. Slightly hairy leaves form basal rosette. Welsh poppy foliage deeply lobed; blue poppies have lance-shaped, toothed leaves. Some blue poppies die after flowering—a little too delicate for some gardeners.

CULTIVATION

Shade to part shade and well-drained soil. Blue poppies need constantly moist soil. Keep blue poppies from blooming the first year, so that they can gain strength; save seeds of blue poppies to propagate. Find them listed as Fertile Blue group or Infertile Blue group. Welsh poppies carefree; will rebloom in fall when rains begin. Scatter seeds of Welsh poppies wherever you want them to grow.

GARDEN USE

A soft look for the woodland garden, where blue poppies should be kept within reach so they can be coddled. Welsh poppies can be left to their own devices to come up here and there, including cracks in pavements.

NOTES

Several species and collections of blue poppies are available by seed.

SPECIES

M. betonicifolia (Himalayan blue poppy)—basal rosette of hairy, heart-shaped leaves; clear blue flowers. Blooms early spring. 4 ft. x 1.5 ft. Zones 7–8.
M. cambrica (Welsh poppy)—lovely and wildflower-like; long, lobed leaves; orange sherbet or lemon ice flowers; color will reseed true in your garden. Blooms midspring to fall. 18 in. x 10 in. Zones 6–8.

Meconopsis betonicifolia

Mimulus aurantiacus

Mimulus

(monkey flower) *Scrophulariaceae*

Flowers have pansylike face, lipped and lobed, but with jagged edges. Foliage and stems tend to have a sticky feel. Often bought as a summer annual, but these species are perennial, although possibly short-lived in the garden.

CULTIVATION

Sun to part shade and evenly moist soil. Shear plants back after bloom for another flowering.

GARDEN USE

Combine with *Lysimachia* in the damp, shady border.

SPECIES

M. aurantiacus (bush monkey flower)—shrubby, short-lived; hybrids and strains in shades of orange and yellow. Blooms spring. 36 in. x 36 in. Zones 7–10. *M. cardinalis* (scarlet monkey flower)—creeping plant; scarlet flowers. Blooms all summer. 36 in. x 24 in. Zones 6–9.

Miscanthus sinensis

(silver grass) *Poaceae*

Statuesque grass for mixed borders or large pots with feathery plumes of pink to purple flowers held above foliage in midsummer. Although herbaceous, it holds up well into midwinter. Fabulous form, great fall color in warm tones.

CULTIVATION

Full sun and well-drained soil. Cut down in late winter as close to ground level as possible. Zones 5–8.

GARDEN USE

Mixes well in perennial combinations among hardy geraniums and asters. Good with small shrubs, such as hebes and dwarf conifers.

CULTIVARS

'Adagio' 3–4 ft. x 3 ft. 'Cosmopolitan' wide leaves with white stripes; 8 ft. x. 4 ft. 'Morning Light' thin white edge to narrow leaves; 5 ft. x 3 ft. 'Silberfeder' pink flower heads fade to white; 7 ft. x 3 ft. 'Yaku Jima' 3–4 ft. x 3 ft. 'Zebrinus' yellow bars on wide leaves give the impression of horizontal stripes; 8 ft. x 3 ft.

Miscanthus sinensis 'Morning Light"

Molinia caerulea 'Variegata'

Molinia caerulea

(Moor grass) *Poaceae*

Herbaceous grass with tidy habit and upright flower stalks in summer.

CULTIVATION

Sun and well-drained soil. Cut down in winter. Zones 4–8.

GARDEN USE

Fine texture works well in the garden or pots, and will not overwhelm neighbors.

SPECIES AND CULTIVARS

M. arundinacea 'Skyracer'—5-ft. flower stems shoot up above 2-ft. foliage.
M. caerulea subsp. *caerulea*—'Moorflamme' dark flowers; 2 ft. x 2 ft. 'Variegata' yellow-striped foliage; 3 ft. x 3 ft.

Monarda

(beebalm, bergamot) *Lamiaceae*

Wild-looking, sometimes two-tiered flowers on tall stems; looks like there's been a petal explosion. Each petal actually a thin, tubular flower on its own. Pinks and reds the predominant flower colors. Like many members of the mint family, has square stems. Leaves have a rough feel.

CULTIVATION

Sun and moist, well-drained soil. Beset with powdery mildew; even those cultivars touted as resistant get it in Pacific Northwest gardens. Many gardeners have given up trying to grow it; others find a modicum of prevention (providing air circulation and consistent moisture—even boggy soil) is all that's needed for a summer full of fabulous flowers. You pay your money and you take your chances.

GARDEN USE

Blooms in midsummer. Enjoy in the summer garden with other plants that need moisture, such as *Lysimachia* and *Mimulus*.

Monarda 'Marshall's Delight'

NOTES

Known as bergamot because it smells similar to the plant used in Earl Grey tea, but it isn't the same.

SPECIES

M. didyma—bushy plant with a leafy base; scarlet flowers. Blooms midsummer to early fall. 36 in. x 24 in. Zones 4–8.

CULTIVARS

(All to about 30 in. x 30 in. unless noted otherwise) 'Coral Reef' coral; 30 in. x 15 in. 'Gardenview Scarlet' red. 'Jacob Cline' deep red; mildew resistant. 'Mahogany' wine-red, dark bracts just below flower. 'Marshall's Delight' pink. 'Petite Wonder' pink; 12 in. x 12 in. 'Prarienacht'/'Prairie Night' lilac. 'Raspberry Wine' deep pink.

Myosotis sylvatica

(forget-me-not) *Boraginaceae*

Sky-blue flowers on lax stems. Lance-shaped, light green leaves with rounded ends.

CULTIVATION

Sun or part shade and well-drained soil. Loves to spread around the garden, and then the flowers fade, quickly followed by powdery mildew. But by that time, we've had enough of them. Cut back hard or pull out and let reseeding take place. Zones 5–9.

GARDEN USE

Blooms spring to early summer. Looks lovely coating the garden floor in spring. Let it grow under shrubs where it's easy to pull up. 12 in. x 6 in.

Myosotis sylvatica

Nectaroscordum siculum

Nectaroscordum siculum

(nectaroscordum) *Alliaceae*

Once included in the onion genus (Allium), this relative grows narrow basal leaves and tall, elegant flower stems. Instead of a ball of flowers, like an onion, it begins with flowers sheathed, like praying hands, and when the sheath splits, out come 1-in., bell-shaped flowers to hang gracefully on 3-ft. stems. Each white flower has a green stripe and is flushed heavily with purple. Like a really cool fireworks display.

CULTIVATION

Full sun and well-drained soil. Zones 5–8.

GARDEN USE

Blooms late spring to early summer. Let it rise above mounds of hebes, geraniums, and lady's mantle.

Nepeta

(catmint) *Lamiaceae*

Pleasing shrubby perennial with gray-green, ridged foliage and square, mint-family stems. Blue flowers, except as noted, are small and tubular. Not true catnip, and so should not be as attractive to cats (but you always hear stories). Some confusion as to where cultivars belong, so don't be surprised to see them mismatched on labels.

CULTIVATION

Full sun and well-drained soil. Low-water-use plant once established; requires no summer water unless noted below. Cutting back spent flower stems encourages another flush of blooms. Cut back whole plant during winter.

GARDEN USE

Small *Nepetas* look good at edge of paths, while more substantial plants should be placed farther back in the border. Grow with garden penstemon, salvias, and hardy geraniums.

SPECIES

N. x *faassenii*—gray-green scalloped leaves; loose spikes of lavender-blue flowers; clump-forming sterile hybrid. Blooms early summer to fall. 18 in. x 18 in. Zones 4–8. 'Kit Kat' dark blue; 15 in. x 15 in.

N. grandiflora—just because its name says "grand" doesn't mean flowers are huge; somewhat sprawling. Blooms early summer; cut back after bloom for more flowers. 30 in. x 30 in. Zones 4–8. 'Dawn to Dusk' light pink against darker bracts.

N. racemosa—'Blue Wonder' 12 in. x 12 in. 'Little Titch' dense mat; blue flowers on tip; 10 in. x 15 in. 'Walker's Low' blue, long-blooming; 24 in. x 24 in.

N. sibirica—10-in. spikes of purple flowers. Blooms mid- to late summer. 36 in. x 24 in. Zones 3–8. 'Souvenir d'Andre Chaudron' compact; blue flowers; 18 in. x 18 in.

CULTIVARS

'Dropmore' lavender blue; 24 in. x 24 in. 'Six Hills Giant' blue flowers, quite impressive; 30 in. x 36 in.

Nepeta 'Six Hills Giant'

Oenothera speciosa

Oenothera

(evening primrose, sundrops) *Onagraceae*

Delicate, large, cup-shaped flowers, some fragrant, that turn up or outward. Some selections open late in the day (hence the name "evening"), but others open earlier in the day. Closed buds, stems, and seedheads can all be ornamental.

CULTIVATION

Full sun and well-drained soil.

GARDEN USE

Grown among other plants, bare bottom stems can be disguised and lax plants find some support.

SPECIES

O. fruticosa (sundrops)—red stems, lance-shaped leaves; yellow flowers in clusters open daytime. Blooms late spring to late summer. 36 in. x 12 in.
Zones 4–8. 'Fyreverkeri'/'Fireworks' red buds, tinted purple foliage. Subsp. *glauca* (syn. *O. tetragona*) red stems, broader leaves tinted red when new; light yellow flowers. 'Sonnenwende'/'Summer Solstice' bright red foliage in summer.

O. macrocarpa (syn. *O. missourensis*)—sprawling, spreading; red buds open late in day to large yellow flowers. Blooms summer. 12 in. x 18 in. Zones 4–8.

O. speciosa—spreading; red stems; fragrant pink flowers. Blooms early summer. 12 in. x 24 in. Zones 5–8. 'Siskiyou' light pink.

CULTIVAR

'Cold Crick' narrow leaves; yellow flowers face up; blooms early summer; 12 in. x 12 in.; zones 5–8.

Omphalodes 'Parisian Skies'

Omphalodes

(navelwort) *Boraginaceae*

Small plants with intense blue flowers above basal foliage.

CULTIVATION

Shade and evenly moist, well-drained soil.

GARDEN USE

Bright plants for the woodland. Set them against *Vancouveria planipetala*, lungworts, and small ferns.

SPECIES

O. cappadocica—grows in a clump. Blooms early to midspring. 10 in. x 12 in. Zones 6–8. 'Cherry Ingram' deep blue; 10 in. x 15 in. 'Parisian Skies' sky blue. 'Starry Eyes' white petal margins.

O. verna—trailing, can be evergreen; flowers blue with white eye. Blooms spring to early summer. 8 in. x 12 in. Zones 6–9.

Origanum

(ornamental oregano) *Lamiaceae*

These may not be fit for the kitchen, but they look lovely in the garden. Flowers are more hoplike than culinary oregano. Foliage usually small, oblong to oval.

Origanum 'Kent Beauty'

CULTIVATION

Full sun and sharp-draining soil.

GARDEN USE

Plant small ones at eye level—at the top of a retaining wall, for example. Grow along with salvias and sun roses.

SPECIES

O. dictamnus (Dittany of Crete)—mounding, trailing; wiry stems; purplish, leafy bracts. Blooms mid- to late summer. 12 in. x 12 in. Zones 8–10.

O. laevigatum—subshrub; purple-pink, papery bracts on upright stems. Blooms late spring to fall. 24 in. x 18 in. Zones 7–10. 'Herrenhausen' more compact flower heads. 'Hopleys' ('Hopleys Purple') dense heads of flowers.

O. rotundifolium—small, clump-forming; wiry stems; leaves clasp stems; hoplike pink bracts in pendulant clusters. Blooms all summer. 12 in. x 12 in. Zones 7–9.

CULTIVAR

'Kent Beauty' more floriferous than *O. rotundifolium*; a delight dangling over edge of a hot wall; 4 in. x 8 in. 'Rosenkuppel' clusters of small pink flowers above foliage, midsummer to fall. 24 in. x 24 in.

Paeonia 'Festiva Maxima'

Paeonia

(peony) *Paeoniaceae*

Big, voluptuous flowers speak of garden bounty. Fragrant single or double flowers are available, as are selections with wad of petals in center of outer petals (reminiscent of those Kleenex flowers we used to make as kids). Flowers in white and shades of pink and red. Foliage usually round-lobed; stems red.

CULTIVATION

Sun and well-drained soil. Plant peony eyes (red growths on roots) no more than 2 in. belowground. Newly planted peony may take several years to settle in before blooming; enjoy the foliage. Susceptible to botrytis; mulch of lime chips (from a stonemason) helps keep spores from splashing up on the plant. Blooms in June, opening—with incredibly bad timing, as far as we are concerned—just before a heavy rainstorm; double flowers fill with water and go crashing to the ground. It's an old story. Single flowers are less likely to collapse (staking is an option, but put stakes in before plants are full-grown). Zones 3–8.

GARDEN USE

Fills the late-spring garden with bounty. Use with roses and near late asters, which can take over in the flower department. Healthy peony foliage turns attractive shades of bronze in fall.

SPECIES AND CULTIVARS

P. lactiflora—blooms late spring to early summer. 24 in. x 30 in. Scores of cultivars or crosses are divided by flower type. Here are a few. Single: 'Blaze' red. Japanese and anemone, with a center mass like shredded petals: 'Bowl of Beauty' pink. Semidouble: 'Coral Charm'. Double: 'Festiva Maxima' white with raspberry flecks. 'Kansas' red. 'Sarah Bernhardt' pink.

P. mlokosewitschii (molly the witch)—single yellow flowers make this an unusual herbaceous peony. Blooms midspring. 36 in. x 24 in.

P. tenuifolia—ferny foliage; deep red to purple-red flowers. Blooms spring. 30 in. x 24 in.

Papaver

(poppy) *Papaveraceae*

Basal foliage, often rough or hairy, sets the stage for flower stems. Cup-shaped, crepe-paper-like flowers are poppies' appeal; some are short-lived but always put on a good show.

CULTIVATION

Full sun and well-drained soil. Smaller poppies can fade away on their own after flowering. Most reseed freely. Oriental poppies can be cut to the ground after flowering.

GARDEN USE

Good for spring and summer flowering. Oriental poppies need some coverage from billowing neighbors; otherwise, their large, ragged rosette of hairy foliage becomes an eyesore after flowering. Seedheads can be left in the garden for ornament and structure.

SPECIES AND CULTIVARS

P. nudicaule (Iceland poppy)—blue-green, divided leaves form rosettes; cup-shaped flowers. Blooms summer. 12 in. x 6 in. Zones 2–8. Champagne Bubbles group apricot, cream, and yellow. Wonderland series pink, coral, and yellow.

P. orientale (Oriental poppy)—bristly leaves; flamboyant, showy, crepe-paper flowers in doubles or singles have noticeable dark spot at base of petals and prominent boss of black stamen. Blooms mid- to late summer. 36 in. x 36 in. Zones 4–9. 'Allegro' scarlet; 16 in. x 16 in. 'Beauty of Livermore' bright red. 'Brilliant' scarlet red. 'Patty's Plum' purple pink. 'Picotee' white with orange edging. 'Royal Wedding' white satin with dark base. 'Turkenlouis'/'Turkish Delight' fringed, fiery red.

Papaver orientale 'Beauty of Livermore'

Penstemon

(beardtongue) *Scrophulariaceae*

Tubular flowers much like foxgloves, attached to upright stems held above leafy bases, appear in late spring or early summer. Leaves are lance-shaped. Although there are many species (some of them shrubs), not all are easy in the mixed garden.

CULTIVATION

Full sun and well-drained soil. Will bloom a second time if spent flower stems are cut back to new growth. After a hard winter, cut back hard for new growth. Short-lived, some cultivars barely make it over the three-year mark of a perennial.

GARDEN USE

Fabulous, long-blooming plant of good, fine texture. Combine with other sun-loving plants such as catmint, lavender, sedum, and ornamental grasses.

SPECIES

P. barbatus—scarlet flowers. Blooms early summer to fall. 36 in. x 24 in. Zones 4–9. Red Riding Hood series pink, purple, blue.

P. digitalis 'Husker Red'—dark purple-black foliage; red flowers. Blooms early to late summer. 20 in. x 18 in. Zones 2–8.

CULTIVARS

Garden hybrids (syn. *P.* x *gloxinioides*) short, linear, evergreen foliage; 36 in. x 36 in., including flower stems. 'Apple Blossom/Thorn' light pink. 'Blackbird' deep purple. 'Garnet' deep wine red. 'Hidcote Pink' deep pink with white throat. 'Margarita BOP' neon blue. 'Midnight' indigo blue. 'Sour Grapes' red-violet.

Penstemon 'Sour Grapes'

Perovskia atriplicifolia

Perovskia atriplicifolia

(Russian sage) *Lamiaceae*

Silver-gray, shrubby plant with sparse foliage. Branched stems and tight spires of deep purple-blue flowers.

CULTIVATION

Full sun and well-drained soil. Sun a must or it flops around. Zones 6–9.

GARDEN USE

Blooms midsummer. An icy cool look for hot sun; looks great with penstemon and salvia. 4 ft. x 3 ft.

CULTIVARS

'Little Spire' to 25 in. 'Longin' more upright, wider leaves.

Persicaria

(fleeceflower, knotweed) *Polygonaceae*

A large genus with a variety of plants. Many have basal leaves and flower stems densely covered in small flowers; others are trailing. Plants have bounced around from genus to genus and may still be listed some places as Polygonum; *some have landed in* Fallopia *or* Tovara.

CULTIVATION

Full sun and well-drained, although moist, soil. Be an informed gardener: Some plants spread, while others form a large clump.

GARDEN USE

Foliage can be coarse, so makes a good foil for finer foliage of ornamental grasses. Shorter selections good for ground covers; taller plants provide color late in the season.

SPECIES AND CULTIVARS

P. affinis (syn. *Polygonum affine*)—spreading plant with leafy base, lance-shaped leaves; stems of pink flowers like thin bottlebrushes. Blooms midsummer to fall. 10 in. x 24 in. Zones 3–8. 'Dimity' 12 in. x 8 in.

P. amplexicaule (bistort)—clumping, forms impressive stand; long oval leaves with a point, leafy base; flower stems rise above, long-blooming. Blooms midsummer to early fall. 4 ft. x 6 ft. Zones 5–8. 'Firetail' crimson; up to 5 ft. x 6 ft. 'Taurus' deep pink.

P. microcephala 'Red Dragon'—dark red leaves with purple, silver, and light-green marking in center; white flowers. Blooms summer. Part shade. 4 ft. x 4 ft. Zones 5–8.

P. virginiana 'Painter's Palette'—foliage brushed white with a red chevron; thin stems of small red flowers not showy. Blooms late summer. 36 in. x 36 in. Zones 3–9.

Persicaria microcephala 'Red Dragon'

Phlox paniculata 'Bright Eyes'

Phlox

(phlox) *Polemoniaceae*

A collection of plants to fit any spot in the garden, from ground cover to tall accent. Five-petaled flowers appear singly or in clusters and sometimes are fragrant. Leaves are oblong with pointed tips.

CULTIVATION

Full sun to part shade and well-drained soil. Most need full sun in the Northwest (where it's never as hot as where phlox came from); exception is if you live inland, where a little light shade is appreciated by some species. Depends greatly on the species; see below. Summer phlox (*P. paniculata*) known for tendency to get powdery mildew; choose resistant cultivars and always provide good air circulation.

GARDEN USE

Short species can be used as ground covers or for trailing in a clump over a wall. Tall phlox give a vertical element to the garden, long before the late asters begin.

SPECIES AND CULTIVARS

P. divaricata (wild blue phlox)—semi-evergreen; flowers light blue, slightly fragrant. Blooms spring. Sun to part shade. 14 in. x 20 in. Zones 4–8. 'Clouds of Perfume' sky blue, more fragrant. 'Sweet Lilac' lilac-mauve.

P. paniculata—tall, unbranched stems topped with domes of highly fragrant flowers. Blooms summer. Sun. 4 ft. x 3 ft. Zones 4–8. 'Becky Towe' yellow variegated foliage; cherry-blossom pink flowers with dark eye; mildew-resistant. 'Bright Eyes' light pink with red eye. 'David' white; mildew resistant. 'David's Lavender' mildew resistant. 'Franz Schubert' lilac pink with dark eye. 'Mt. Fuji' white. 'Starfire' deep red.

P. stolonifera (creeping phlox)—evergreen, mat-forming, roots at nodes; pink or lilac flowers. Blooms spring. Sun to part shade. 6 in. x 12 in. Zones 4–8. 'Blue Ridge' pale blue.

P. subulata (moss pink)—evergreen mound, stiff needlelike green leaves; star-shaped flowers in pink, purple, or white. Blooms late spring to early summer. Sun; no water in summer. 6 in. x 20 in. Zones 3–8. 'Fort Hill' deep pink, fragrant.

Phormium

(New Zealand flax) *Phormiaceae*

Swordlike evergreen leaves, often striped with white or gold flowers on stalks, appearing inconsistently from year to year, but no matter, they are unnecessary. Dramatic plants for form and texture, adding an exotic note to any planting. Blooms early summer.

Phormium 'Rainbow Sunrise'

CULTIVATION

Full sun and well-drained soil. Zones 8–10.

GARDEN USE

Contrast sweet flowers and mounds of leaves with the sharp (looking), pointy leaves.

NOTES

Remove any dead or damaged leaves by cutting them as close to the ground as possible.

CULTIVARS

'Jack Spratt' dark bronze foliage; 18 in. x 18 in. 'Maori Queen' pink and apricot striped leaves; 5 ft. x 5 ft. 'Rainbow Sunrise' pink leaves with red edge; 3 ft. x 4 ft.

Polemonium

(Jacob's ladder) *Polemoniaceae*

Lovely little clusters of small, open, bell-shaped flow-ers. Each long leaf is divided evenly into short leaflets that look like steps of a ladder.

CULTIVATION

Part shade and well-drained, well-mulched soil. Cut back hard after it blooms to encourage a fresh set of new leaves. Zones 4–9.

GARDEN USE

Blooms early summer. A sweet plant for the partly shady garden. Foliage offers an interesting texture.

SPECIES AND CULTIVARS

P. caeruleum—blue flowers. 24 in. x 12 in. 'Brise d'Anjou'/'Blanjou' heavily variegated leaves edged in cream. 'Snow and Sapphires' white-edged leaves; hardy to Zone 3.

P. carneum—apricot-pink flowers, slightly fragrant. 16 in. x 8 in. 'Apricot Delight' deeper apricot.

Polemonium caeruleum 'Brise d'Anjou'

Polygonatum biflorum

Polygonatum

(Solomon's seal) *Liliaceae*

Enchanting woodland plant with unbranched, arching stems and horizontally held leaves that clasp stalks. Small, white, pendant bell-shaped flowers appear in leaf axils in ones or twos in late spring to early summer, followed by black or red berries.

CULTIVATION

Shade to part shade and well-drained, humusy soil. Grows from rhizomes, so new single stems emerge to increase the stand.

GARDEN USE

Create a lovely tableau in the garden. Grow with false Solomon's seal, hostas, and *Omphalodes* for a shot of blue.

SPECIES AND CULTIVARS

P. biflorum—arching stems; pairs of small, greenish-white bell flowers hang from leaf axils; dark blue berries in fall; graceful but highly variable in size. 1–5 ft. x 2 ft. Zones 3–9. 'Prince Charming' 12 in. x 36 in.

P. x *hybridum* 'Striatum'—creamy stripes on foliage. 5 ft. x 1 ft. Zones 6–9.

P. odoratum—slightly arching; elegant, dark stems; fragrant flowers in ones or twos. 32 in. x 12 in. Zones 4–8. Var. *pluriflorum* 'Variegatum' (often listed without the variety name) creamy white–edged leaves; dark red stems.

Polystichum setiferum

Polystichum

(Alaska fern, sword fern, tassel fern) *Dryopteridaceae*

Handsome evergreen ferns. Ferns are nonflowering plants.

CULTIVATION

Shade to part shade and well-drained soil. Takes summer dryness. Although evergreen, has a neater appearance if old fronds are cut off in late winter, before new fronds emerge.

GARDEN USE

Good architectural element for the shade garden.

SPECIES AND CULTIVARS

P. munitum (sword fern)—it's everywhere in our woods, but should be in everyone's garden too; an accommodating plant. 36 in. x 36 in. Zones 3–8.

P. polyblepharum (tassel fern)—glossy foliage well-cut, giving a fine texture. 36 in. x 36 in. Zones 6–8.

P. setiferum (Alaska fern)—4 ft. x 3 ft. Zones 6–9. Divisilobum group shaggy look. 'Congestum Cristatum' looks like it has little wads of extra fronds at end of each frond.

Potentilla

(cinquefoil) *Rosaceae*

Strawberry-like leaves divided into three leaflets; serrated with a quilted texture. Herbaceous potentillas (shrubs not listed here) have mostly basal foliage. Five-petaled flowers about 1 in. across and have an open face.

CULTIVATION

Full sun and well-drained soil. Once established, does well without summer water. Zones 5–8.

GARDEN USE

Blooms early to midsummer, some cultivars continuing. Long flower stems weave among other plants, creating ever-changing vignettes. Best to grow where its slightly bald middle isn't noticeable. Good with roses, hardy geraniums, veronicas, and hebes.

CULTIVARS

'Blazeaway' orange-red. 'Gibson's Scarlet' deep scarlet. 'Melton Fire' large red center with deep, yellow outer rim. 'Miss Willmott' cherry red with dark eye. 'Monarch's Velvet' raspberry-red with large, dark center. 'Ron McBeath' lipstick pink with red center. 'William Rollison' semidouble; orange with yellow markings. 'Yellow Queen' yellow with red eye.

Potentilla 'Miss Willmott'

Primula

Primula

(primrose) *Primulaceae*

There's more than just the grocery store variety—not that there's anything wrong with those; but this is a huge genus divided and subdivided into various groups. Although many of these specialty plants are delicate and choosy, there are still many that are good garden plants. Basal rosette of wide, spoon-shaped leaves and flower stems rise out of this. Bell-shaped or flat flowers can appear singly or in clusters.

CULTIVATION

Shade to part shade and well-drained, evenly moist soil. Candelabra type for streamside, wet soils; auriculas take less water.

GARDEN USE

Tall candelabra types look stunning growing in a swath in a wet depression in the garden. Smaller primroses need to be up close and personal to be appreciated. Auriculas grow best in pots, as they can get lost in the garden (or drown in winter wet).

NOTES

Control slugs.

SPECIES

P. auricula—almost succulent foliage in tight rosettes; flowers look like a Victorian print, open-faced, rounded petals, often two-toned, some with "meal" (looks like little spots of flour). Blooms early to midspring. Part shade. 10 in. x 5 in. Zones 3–8. Named cultivars exist, but we mostly see those auriculas available only from specialty societies; generally, find mixed seed lots at nurseries in spring and buy in bloom.

P. beesiana, *P. bulleyana*, *P. japonica*, and *P. pulverulenta* are candelabra types—basal leaves; tall, elegant stems, flowers in yellow, pink, rose, or orange a beautiful sight. Blooms spring. Part shade; will reseed, to your delight. 24 in. x 18 in. Zones 3–8.

P. marginata—usually included in auricula group; tiny; encrusted leaf edges; small stems of lavender flowers. Blooms early spring. Part shade. 6 in. x 6 in. Zones 3–8.

P. vialli—an oddity, with unusual two-toned cones of flowers; red cones open to lavender from bottom up. Blooms late spring to early summer. Shade to part shade. 24 in. x 24 in. Zones 5–8.

CULTIVARS

Of both *P. polyanthus* and *P. vulgaris*: 'April Rose' double red-pink. 'Green Lace' green frilled flowers with yellow throat. 'Guinevere' bronze foliage; single pink flower with yellow eye. 'Miss Indigo' double deep violet. Blooms spring. Part shade. 8 in. x 8 in. Zones 5–9.

Prosartes—see *Disporum*

Pulmonaria 'Roy Davidson'

Pulmonaria

(lungwort, soldiers-and-sailors) *Boraginaceae*

Small mounding or trailing herbaceous plants for shade to part shade; all bloom late winter and early spring. Clusters of small, bell-shaped flowers open in early to midspring, often changing from pink in bud to blue in flower. Oblong leaves often spotted with silver.

CULTIVATION

Shade and well-drained, moderately moist soil; provide a good mulch. Not drought tolerant. After flowering, plagued by powdery mildew; keep it at bay by cutting back hard after flowers have finished. Then a flush of new growth will emerge (although some of the selections are so covered in silver-white spots that it looks like they have powdery mildew anyway). Zones 5–8.

GARDEN USE

Delightful in the shade garden, where it brings spring color early. Combine with small *Carex conica* 'Snowline', *Brunnera* 'Jack Frost', and some solid green hostas.

SPECIES AND CULTIVARS

P. angustifolia—long, unspotted leaves; clusters of pink buds open to blue flowers. Blooms late winter to early spring. 12 in. x 18 in.

P. longifolia—spotted long leaves; flowers violet blue. 12 in. x 18 in. Subsp. *cevennensis* longer leaves, more disease resistant.

P. officinalis—spotted leaves; pink buds to violet-blue flowers. 12 in. x 18 in.

P. rubra (syn. *P. montana*)—unspotted leaves; coral-red flowers. 12 in. x 36 in. 'David Ward' leaves edged in cream. 'Redstart' brick-red flowers.

P. saccharata (Bethlehem sage)—leaves heavily spotted; pink buds open to blue flowers. 12 in. x 24 in.

CULTIVARS

'Benediction' lightly speckled foliage; deep blue flowers. 'Berries and Cream' spotted; raspberry pink flowers. 'Majeste' solid leaves, silver; pink to blue flowers. 'Roy Davidson' spotted; pink to blue flowers. 'Raspberry Splash' spotted; coral pink flowers. 'Sissinghurst White' spotted; white flowers. 'Trevi Fountain' narrow leaves; dark blue flowers.

Pulsatilla vulgaris

Pulsatilla vulgaris

(syn. Anemone pulsatilla; pasque flower)
Ranunculaceae

Softly hairy foliage, deeply divided, with pink to carmine flowers that show a cluster of yellow-orange stamen. Showy, silky seedheads. Blooms in spring, around Eastertime.

CULTIVATION

Full sun and well-drained soil. Dormant in summer. 8 in. x 8 in. Zones 5–7.

GARDEN USE

Plant near late-emerging perennials such as hardy geraniums that can distract the eye.

CULTIVARS

'Alba' white flowers. Var. *rubra* deep red.

Rodgersia

(rodgersia) *Saxifragaceae*

Wonderfully architectural herbaceous plant that forms sizable clumps. Large, compound leaves have a bit of texture. Tall, dark flower stems emerge in summer, covered in a loose cone shape with tiny pink or white flowers.

CULTIVATION

Shade to part shade and constantly moist soil, even streamside. Here's a trick to keeping the soil wetter for Rodgersia than neighboring plants: At planting time, place a wide, flimsy plastic plant saucer under the root ball. This will catch and hold water for the Rodgersia but let the rest of the soil in the bed drain well. Zones 5–8.

GARDEN USE

Blooms summer. Great effect in the shade garden, along with other water-lovers such as astilbe. Give it room to spread its leaves.

Rodgersia pinnata

SPECIES AND CULTIVARS

R. aesculifolia—round, divided leaves up to 10 in. across resemble leaves of horse chestnut tree (*Aesculus*); stalks of fragrant white or pink flowers. 3–5 ft. x 3 ft.

R. pinnata—bronze-tinted foliage, divided leaves shaped like a feather, up to 8 in. across, leaflets wider at end than stem; red stalks, pink flowers. 4 ft. x 3 ft. 'Elegans' creamy white. 'Superba' long-blooming; bright pink.

R. podophylla—large, fan-shaped leaves, each leaflet up to 10 in. long, bronze-tinted; stalks of creamy white flowers. 5 ft. x 4 ft.

Romneya coulteri

Romneya coulteri

(California tree poppy, Matilija poppy) *Papaveraceae*

Impressively tall, with running roots and blue-green, divided leaves. Flowers are 5 in. across and look like a giant fried egg—white with a big yellow center.

CULTIVATION

Sun and well-drained, poor soil. Zones 7–10.

GARDEN USE

Blooms summer. Plant in a hot, sunny, dry place edged by concrete (such as sidewalks or driveways) to help contain it. 8 ft. x indef.

Rudbeckia

(black-eyed Susan, coneflower) *Asteraceae*

Flowers big yellow or gold daisies, usually with dark centers, held above basal foliage on sparsely foliated stems.

CULTIVATION

Full sun and well-drained soil. Deadheading prolongs flowering. Tallest selections may need staking or support of neighboring plants. Zones 5–9.

GARDEN USE

Bright summer color for the late-season garden. Looks good with just about anything in late summer, including ornamental grasses. Tall, lanky coneflowers are more than an oddity; they offer vertical zing.

SPECIES AND CULTIVARS

R. fulgida—erect, hairy, dark green leaves; orange 3-in. daisy flowers with dark disk. Blooms mid- to late summer. 36 in. x 18 in. Most common: var. *sullivantii* 'Goldsturm' more floriferous, flowers to 4 in. across; 24 in. x 24 in.

R. laciniata—tall! with hairy foliage, egg-shaped leaves, branching stems; flowers golden yellow with drooping petals. Blooms mid- to late summer. 10 ft. x 3 ft.

R. occidentalis 'Black Beauty'—all dark brown cone, no petals, just a green, leafy, calyx collar; odd, but finches like the seed. Blooms summer. 36 in. x 24 in. 'Green Wizard' similar.

Rudbeckia fulgida var. *sullivantii* 'Goldsturm'

Ruta graveolens 'Jackman's Blue'

Ruta graveolens

(rue) *Rutaceae*

Blue-green, ferny foliage. Clusters of small yellow flowers appear in summer.

CULTIVATION

Sun and well-drained soil. Zones 5–9.

GARDEN USE

Foliage makes an exceptionally interesting look in the garden when combined with bright spiky flowers of *Veronica* and midseason mound of *Chrysanthemum* 'Clara Curtis'. 36 in. x 30 in.

NOTES

Plant's oil can cause blistering on skin, especially in sun, so be careful when cutting back.

CULTIVARS

'Jackman's Blue' foliage more blue.

Salvia greggii

Salvia

(sage) *Lamiaceae*

The genus Salvia includes many small shrubs, in addition to culinary sage (not described here). Only a few of the many perennials, and shrubs that we treat as perennials, are noted. Opposite leaves and lipped and lobed flowers, as is common in the mint family. Selections bloom in spring or summer; flowers vary in color and size.

CULTIVATION

Full sun and well-drained soil. Wet winter soil will kill more salvias than cold, although most also need heat. All should be cut back in spring, even the rather shrubby *S. greggii* cultivars—for those, watch to see where new growth will start on the bare stems, and cut just above that. Cutting earlier can lead to salvia death.

GARDEN USE

Include in the hottest part of the garden. Small, shrubby salvias do well in containers; larger, more lax species can flop around on a hillside. Great summer color; makes us think we live in a desert.

NOTES

Good hummingbird plants.

SPECIES

S. greggii (autumn sage)—small, glossy green leaves on shrubby plant; colorful, lipped, bright red flowers. Blooms summer to fall. 20 in. x 20 in. Zones 7–9. Seedling strains listed as coral, orange, or pink.

S. nemorosa—basal leaves; thickly blooming stems of purple flowers. Blooms summer to fall; cut back for continuous bloom. 36 in. x 24 in. Zones 5–9. 'Ostfriesland'/'East Friesland' intense violet. 'Royal Distinction' rose pink; 16 in. x 18 in.

S. patens (gentian sage)—deep blue flowers, with beaklike appearance, in leaf axils. Blooms midsummer to fall. Spreads by tuberous roots; mulch with sword-fern fronds over winter. 24 in. x 18 in. Zones 8–9.

S. x *superba* (or 'Superba')—clump-forming; violet flowers. Blooms early to midsummer. 32 in. x 12 in. Zones 5–9.

S. x *sylvestris*—18 in. x 18 in. Zones 5–9. 'Blauhugel'/'Blue Hill' light blue. 'Blaukonigin'/'Blue Queen' purple-blue. 'Mainacht'/'May Night' violet-blue. 'Rosakonigin'/'Rose Queen' rose.

S. verticillata 'Purple Rain'—large leaves with ruffled edges; deep purple flowers in spikes. Blooms summer. 24 in. x 24 in. Zones 6–10.

CULTIVAR

'Hot Lips' red flowers with white center; blooms spring into summer; 3 ft. x 3 ft. 'Indigo Spires' basal leaves; intense violet-blue flowers; blooms summer; 36 in. x 12 in.; zones 7–9.

Sanguisorba

(burnet) *Rosaceae*

A wildflower look about it. Basal foliage and long leaves divided and toothed. Oblong or egg-shaped clusters of tiny flowers on tall, wiry stems.

CULTIVATION

Full sun and moist, well-drained or constantly wet soil. Zones 4–8.

GARDEN USE

Blooms late summer. Airy texture, late flowering work well with the naturalistic style.

SPECIES

S. canadensis (American burnet)—white flowers in elongated clusters up to 5 in. long. Blooms late summer to early fall. 4 ft. x 2 ft.
S. menziesii—Northwest native; blue-green foliage; red-purple flowers beginning late spring. 30 in. x 18 in.
S. obtusa (Japanese burnet)—arching red spikes of flowers. 24 in. x 24 in.
S. officinalis—purple-brown flowers. Blooms early to late summer. 4 ft. x 2 ft. 'Lemon Splash' yellow spots on foliage. 'Red Thunder' deep red.

Sanguisorba canadensis

Saponaria ocymoides

Saponaria

(soapwort) *Caryophyllaceae*

Lance-shaped or oval leaves. Single, small, open-faced flowers.

CULTIVATION

Sun and well-drained soil. Shear back after initial bloom to encourage another show.

GARDEN USE

Frothy, airy mounds give sunny spaces in the garden a fine texture. Grows well at the top or bottom of a wall, below a birdbath, or at corners of paths.

SPECIES AND CULTIVARS

S. x *lempergii* 'Max Frei'—lance-shaped leaves; red flowers. Blooms early to midsummer. 12 in. x 18 in. Zones 5–8.

S. ocymoides (rock soapwort)—mass of deep pink flowers. Blooms early summer. 6 in. x 18 in. Zones 4–8. 'Snow Tip' white flowers.

S. officinalis (bouncing bet)—light pink flowers. Blooms summer to fall. Pinch back early growth for bushier plant. 24 in. x 20 in. Zones 3–9. 'Rosa Plena' deep pink; double.

Saxifraga

(saxifrage) *Saxifragaceae*

One of the messier genera, taxonomically speaking: It's easier to look for the descriptive common name "mossy saxifrages" rather than a botanical name. Tight, cushiony clumps of finely cut evergreen foliage. Sprays of white, pink, or red flowers held above foliage on thin stems.

Mossy saxifrage

CULTIVATION

Full sun and well-drained soil. Tends to rot out in the middle over time. Zones 6–8.

GARDEN USE

Blooms early spring. Specialty rock-garden plants and several all-around good garden plants.

SPECIES AND CULTIVARS

S. stolonifera (strawberry saxifrage, mother of thousands)—evergreen; thick, rounded, dark green leaves with white veins, slightly hairy with red underside; red runners look like threads dropping to the ground; sprays of white flowers. Blooms summer. Good for rockeries and baskets. 12 in. x 12 in. Zones 6–9.

S. x *urbium* (syn. *S. umbrosa*; London pride)—succulent, evergreen leaves shaped like spatulas; sprays of pink flowers on wiry stems. Blooms early summer. Great for dry shade; increases by rosettes. 12 in. x indef. Zones 6–8. 'Pumiloides' smaller in all parts.

Scabiosa

(pincushion flower) *Dipsacaceae*

A cottage garden plant with a soft look and frilly, girlish flowers. Flowers an outer ring of ruffled petals and often another set of shorter petals in the middle, with showy cluster of stamen (pins in a cushion, just as the name says).

CULTIVATION

Full sun and well-drained soil. Many have a long blooming period, but it always helps to keep them deadheaded.

GARDEN USE

Grow with contrasting shapes such as crocosmia and *Kniphofia*. Looks as wonderful in a vase as in the garden, so keep it within reach.

SPECIES

S. caucasica—lance-shaped leaves that divide; 2-in.-wide, flat, light blue flowers with ruffled outer petals and center of short, gathered petals with stamen protruding. Blooms midsummer. 24 in. x 24 in. Zones 4–9. 'Fama' deep lavender-blue with white center, 3 in. across.

S. columbaria—base of egg-shaped leaves; tall stems; light lavender-blue, 1.5-in. flowers. Blooms summer to early fall. 28 in. x 36 in. Zones 3–8. 'Nana' pink; to 10 in. Subsp. *ochroleuca* creamy yellow flowers on thin stems; 30 in. x 20 in.

CULTIVARS

'Butterfly Blue' pale green foliage; light blue flowers; blooms spring to fall over a long period; 12 in. high. 'Pink Mist' 2-in. pink flowers with pale center; long-blooming; 16 in. x 16 in.

Scabiosa caucasica 'Fama'

Schizostylis coccinea
'Oregon Sunset'

Schizostylis coccinea

(syn. Hesperantha; crimson flag) *Iridaceae*

Showy pink 1-in., star-shaped flowers, several on a stem. Swordlike foliage.

CULTIVATION

Sun or part shade and well-drained to submerged soil. Grows from rhizomes that spread, especially in moist soil, to form a substantial colony. Zones 7–9.

GARDEN USE

Good with dampish partners such as *Lysimachia*. 24 in. x 12 in.

SPECIES AND CULTIVARS

F. alba white flowers. 'Oregon Sunset' pink-red. 'Sunrise' deep pink.

Sedum

(stonecrop) *Crassulaceae*

Fabulous succulent for everything from ground cover to tall, flowery component; some are evergreen, others deciduous. Foliage often tinted or variegated, so even if flowers are not the most ornamental feature, it's still worthy of a space in the garden.

CULTIVATION

Full sun and well-drained soil. Most take the dry weeks of summer well, but some taller varieties do better if given a little supplemental water then. 'Herbstfreude'/'Autumn Joy' and other tall sedums tend to spread out in the middle once they reach their full height; to prevent this, cut stems back by about one-third in midspring or grow in full sun and poor soil with no supplemental water—that will make them stand up.

GARDEN USE

Use in hot, sunny spots of the garden—parking strips, rockeries, pathways. Tall plants look wonderful combined with penstemon, catmint, and bronze sedges. Use with salvias, too.

SPECIES

S. cauticola—deciduous; small, gray leaves with purple spots, foliage forms tiny rosettes along stem; rose-pink flowers. Blooms early fall. 3 in. x 12 in. Zones 5–9. 'Lidakense' more compact. 'Robustrum' larger foliage, longer stems; rose-red flowers.

S. erythrostictum 'Frosty Morn'—deciduous; mint-green foliage heavily edged in white; white or pale pink loose clusters of flowers. Blooms late summer. 12 in. x 24 in. Zones 6–9.

S. kamtschaticum—spreading plant; loose, short clusters of yellow flowers age red. Blooms fall. 4 in. x 8 in. Zones 6–9. Var. *kamtschaticum* 'Variegatum' thin edge of cream makes green leaves stand out.

S. sieboldii—evergreen; round leaves on stems expanding from center; bright pink flowers. Blooms fall. 4 in. x 8 in. Zones 6–9. 'Variegatum' green-cream-pink.

S. spathulifolium 'Purpureum'—purple, evergreen foliage; mat-forming; short stems of yellow flowers. Blooms summer. Good for little hot spots in rockeries. 4 in. x 12 in. Zones 5–9.

S. spectabile—deciduous; upright, fleshy stems with thick, round, pale green leaves; flat clusters of pale pink flowers. Blooms late summer. 18 in. x 18 in. Zones 4–9. 'Brilliant' deep pink. 'Meteor' big,

Sedum telephium subsp. *ruprechtii*

rose-pink clusters. 'Neon' larger, magenta flower heads. 'Stardust' white.

S. spurium—semi-evergreen; mat-forming; hairy red stems, toothed leaves; flowers purple with orange center on short stems. Blooms late summer. 4 in. x 24 in. Zones 4–9. 'Schorbuser Blut'/'Dragon's Blood' red flowers; bronze foliage. 'John Creech' small green rosettes of foliage; 2 in. x spreading.

S. telephium—tuberous rooted, clump-forming; thick stems; fleshy, egg-shaped leaves; purple-pink flower clusters. Blooms late summer. 24 in. x 12 in. Zones 4–9. 'Matrona' stems purple-red. 'Mohrchen' burgundy leaves turn red in fall. Subsp. *ruprechtii* dusky foliage.

CULTIVARS

'Bertram Anderson' purple-blue foliage; deep pink flowers bloom late summer; 15 in. x 12 in. 'Herbstfreude'/'Autumn Joy' green foliage; blooms late summer, crimson flower heads age brown; 24 in. x 24 in. 'Purple Emperor' deep purple foliage; pink flowers; 18 in. x 18 in. 'Ruby Glow' purple-flushed leaves; red flowers; 10 in. x 18 in. 'Vera Jameson' deep purple foliage; short flower stems, pink flowers; 12 in. x 18 in.

Sidalcea

(checkerbloom, false mallow) *Malvaceae*

Clump-forming plants with round, lobed leaves and hollyhock-like flowers.

CULTIVATION

Full sun and well-drained soil. Zones 6–8.

GARDEN USE

A good, albeit short, substitute for hollyhocks in the rust-prone garden.

SPECIES

S. candida—Round, coarsely lobed basal leaves; spikes of small, light blue flowers. Blooms mid- to late summer. 32 in. x 18 in. 'Bianca' white.
S. malviflora (prairie mallow, checkerbloom)—round, shallowly lobed basal leaves; spikes of 2-in.-wide pink flowers. Blooms early to midsummer. 4 ft. x 1.5 ft.

CULTIVARS

'Elsie Heugh' pale pink, fringed petals. 'Party Girl' 3-in., deep pink flowers; 36 in.

Sidalcea 'Party Girl'

Silene uniflora 'Druett's Variegated'

Silene

(campion, catchfly) *Caryophyllaceae*

Upright or mat-forming plants with five-petaled flowers that can resemble little pinwheels. Some have what looks like inflated bladders at base of flower.

CULTIVATION
Sun and well-drained soil.

GARDEN USE
Blooms summer. Grow with ornamental onions or contrasting colors of sneezeweed (**Helenium**) and **Heliopsis**.

SPECIES AND CULTIVARS
S. dioica—unassuming wildflower; upright plant; a few rose-pink flowers atop stems. Blooms late spring to midsummer. Sun to part shade. 32 in. x 18 in. Zones 6–9. 'Clifford Moore' creamy-edged leaf margins.
S. uniflora (sea campion)—small plant; pink flowers resemble single carnations. Blooms summer. 6 in. x 8 in. Zones 3–7. 'Druett's Variegated' creamy-edged leaves but can revert to all green; white flowers.

Sisyrinchium

(blue-eyed grass) *Iridaceae*

Comes in colors other than blue, too—yellow and mauve-pink. Some have grassy foliage; others, sword-like iris leaves.

CULTIVATION

Full sun and well-drained soil; most like regular water.

GARDEN USE

Smaller ones get overwhelmed easily—plant them in sight; taller *S. striatum* can hold its own. Grow with sundrops (*Oenothera*), geraniums, and mounding asters or chrysanthemums.

SPECIES

S. angustifolium (blue-eyed grass)—semi-evergreen, grassy foliage; purple-blue flowers. Blooms early summer. 12 in. x 4 in. Zones 5–8.
S. californicum (golden-eyed grass)—semi-evergreen, grassy foliage; bright yellow flowers. Blooms summer. 24 in. x 6 in. Zones 8–9.
S. striatum—clump-forming; irislike foliage; creamy yellow flowers. Blooms early summer. 36 in. x 10 in.

Sisyrinchium striatum

Zones 7–8. 'Aunt May' white variegation up leaves; white flowers.

CULTIVAR

'Quaint and Queer' charming, with mauve-purple flowers; blooms early summer; 12 in. x 12 in.; Zones 6–8. 'Devon Skies' grassy foliage; blue flowers throughout summer; 6 in. x 10 in.; Zones 7–9.

Smilacina

(syn. Maianthemum; false Solomon's seal)
Asparagaceae

Arching stems, oblong leaves with pointed tips clasps the stems. White fragrant flowers followed by red berries. Wonderful woodland plants. Blooms mid- to late spring.

Smilacina racemosa

CULTIVATION

Shade to part shade and well-drained, humusy soil. Zones 4–8.

GARDEN USE

Combine with lily-of-the-valley, hostas, ferns and true Solomon's seal (*Polygonatum*).

SPECIES

S. racemesa (false Solomon's seal)—frothy clusters of tiny flowers at stem end. 36 in. x 24 in.
S. stellata (star flower)—several star-shaped flowers at stem end. 24 in. x 24 in.

Solidago

(goldenrod) *Asteraceae*

Although a member of the daisy family, does not sport usual big daisy flowers. Instead, blooms in sprays of golden or pale yellow on tall or short stems. Oblong to lance-shaped leaves.

CULTIVATION

Full sun and well-drained soil. Cut back in late winter, before new foliage begins growing.

GARDEN USE

Good plant: fine texture and form for the garden at a time—mid- to late summer—when all other flowers seem to be big daisies. Good with late roses, ornamental grasses, dark-leaved bergenias.

NOTES

Has been unfairly slapped with the label of allergy-inducer, but does not cause hayfever.

SPECIES AND CULTIVARS

S. canadensis—stems topped with sprays of golden yellow flowers. Rhizomatous plant for the natural garden. 5 ft. x 1 ft. Zones 3–9.

Solidago rugosa 'Fireworks'

S. rugosa 'Fireworks'—leaves and stems rough to the touch; showy fireworks fun with bright yellow streaks of flowers. 3–4 ft. x 2 ft. Zones 3–9.

S. sphacelata 'Golden Fleece'—wide leaves; bright yellow flower clusters almost flat-topped. Blooms fall. 18 in. x 2 in. Zones 5–9.

CULTIVARS

'Crown of Rays' long sprays of yellow flowers; 24 in. x 18 in. 'Goldkind'/'Golden Baby' almost flat, wide sprays of bright yellow flowers; 20 in. x 24 in.

Stachys

(betony, lamb's ears) *Lamiaceae*

An interesting collection of plants; all have in common the mint family characteristics of lipped, tubular flowers and square stems. From there, they branch out into big and tall, short and stout, lax and upright.

CULTIVATION

Full sun and well-drained soil, except as noted below. Most common garden plant in this genus is lamb's ears (*S. byzantina*). Its lax stems can snake around in the garden, but when you yank them out, you are rewarded with a scent of fresh apples. Who can complain about that?

GARDEN USE

Grow for interesting foliage as well as flowers (or, in some cases, instead of flowers). A good component of the summer mixed garden, along with sedums and hebes.

SPECIES AND CULTIVARS

S. albotomentosa (species is same as cultivar 'Hildago')—light green foliage (smells like 7-Up), white woolly stems, somewhat sprawling; salmon-colored, lipped flowers. Blooms all summer. 16 in. x 20 in. Zones 7–9.

S. byzantina (lamb's ears)—silver-gray, felted leaves as soft indeed as a lamb's ear must be, prostrate stems; upright flower stalks have purple flowers within leaf axils. Blooms early summer. May reseed a bit; loved by bees. 18 in. x 24 in. Zones 4–8. 'Big Ears' (syn. 'Countess Helen von Stein') large leaves. 'Primrose Heron' new foliage yellow-green.

S. coccinea (Texas betony, scarlet hedge nettle)—green, wrinkled foliage; spikes of deep pink-red flowers. Blooms midspring to summer. Put it in the hottest spot you have. 24 in. x 18 in. Zones 7–9.

S. macrantha—clump-forming, dark green, wrinkled, hairy foliage; violet-pink flowers. Blooms early summer to fall. Takes part shade. 24 in. x 12 in. Zones 5–7. 'Rosea' rose-pink. 'Superba' pink.

S. officinalis—basal clump of leaves; violet-pink flowers. Blooms late spring to early summer. 24 in. x 12 in. Zones 5–8.

Stachys byzantina

Stokesia 'Purple Parasols'

Stokesia laevis

(Stoke's aster) *Asteraceae*

Large, cup-shaped, light blue flowers with at least two rows of deeply fringed petals. Flowers sit atop leafless stems above a clump of long, dark, lance-shaped leaves.

CULTIVATION

Full sun and well-drained soil. Takes the dry weeks of summer well. Zones 5–9.

GARDEN USE

Blooms midsummer to fall. Combines well with spiky plants such as veronicas and hebes. 24 in. x 18 in.

CULTIVARS

'Alba' white flushed pale pink. 'Blue Danube' sky blue; 16 in. x 16 in. 'Klaus Jellito' large blue flowers. 'Purple Parasols' big 4-in., purple-violet, ruffled flowers.

Symphytum

(comfrey) *Boraginaceae*

Coarse texture. Leaves are oblong to lance-shaped and hairy. Branches of small, nodding flowers. Some selections more ornamental than the herb garden plant.

CULTIVATION

Full sun to part shade and well-drained soil. A tenacious hold on your garden once you plant it; even small pieces of root will grow. Zones 4–8.

GARDEN USE

Grow with late-summer flowering plants that will complement the foliage, such as late asters and geraniums.

SPECIES AND CULTIVARS

S. x *uplandicum*—gray-green leaves; violet flowers. Blooms late spring to early summer. Part shade. 4 ft. x 3 ft. Zones 3–9. Known mostly by these cultivars: 'Axminster Gold' gold edges to leaves. 'Variegatum' wide, creamy white margins; looks best if flower stems are removed.

Symphytum x *uplandicum* 'Axminster Gold'

Tanacetum

(feverfew, pyrethrum, tansy) *Asteraceae*

An assortment of plants with daisylike flowers. Foliage is well-cut and green or silver.

CULTIVATION

Sun and well-drained soil.

GARDEN USE

Ground covers or cut flowers. Grow with roses, lavender, and other border plants.

NOTES

Should not be confused with the noxious weed tansy ragwort (*Senecio jacobaea*).

SPECIES AND CULTIVARS

T. coccineum (pyrethrum)—bright green, ferny leaves; daisy flowers in shades of red and pink with yellow center. Blooms early summer. 30 in. x 18 in. Zones 5–9.

T. densum subsp. *amani*—forms mat of small, ferny, silver-gray leaves; yellow daisy flowers rather mar the look. Blooms summer. Excellent ground cover for a dry, sunny spot. 10 in. x 8 in. Zones 6–8. 'Beth Chatto' almost identical.

T. parthenium (feverfew)—well-cut, chrysanthemum-like leaves and dark stems; covered in bloom with small, white-petaled, yellow-centered daisies. Blooms summer; blooms again if cut back. Sun to part shade; takes poor circumstances well; reseeds, but makes a great filler. 24 in. x 12 in. Zones 4–9. 'Aureum' yellow foliage. 'Plenum' or 'Flore Pleno' double the petals.

Tanacetum 'Beth Chatto'

Thalictrum rochebruneanum

Thalictrum

(meadow rue) *Ranunculaceae*

Flowers often small but come in big clusters atop soft, green stems full of foliage.

CULTIVATION

Part shade and well-drained, humusy soil. Zones 5–9.

GARDEN USE

Tall plant with summertime flowers gives the woodland garden a little variety in contrast to lots of small, ferny plants that grow in part shade. Adds another layer to the shade garden; plant with hostas, lungworts, *Omphalodes*, and Welsh poppies.

SPECIES AND CULTIVARS

T. aquilegifolium—leaves like a columbine, blue-green and divided twice; clusters of small lavender flowers. Blooms early summer. 36 in. x 18 in. 'Purpureum' dark purple stems and flowers.

T. delavayi—divided leaves; clusters of small lilac flowers. Blooms midsummer. 4 ft. x 2 ft. 'Album' white. 'Hewitt's Double' double flowers.

T. flavum—divided leaves; stalks of pale yellow flowers. Blooms summer. 36 in. x 18 in. 'Illuminator' pale yellow flowers; bright green leaves. Subsp. *glaucum* blue-green foliage.

T. rochebruneanum—divided leaves; dark purple flower stems, blue-green foliage; loose clusters of lavender flowers. Blooms summer. 36 in. x 12 in. 'Black Stocking' dark stems; fluffy magenta flowers. 4 ft. x 2 ft.

Tiarella

(foamflower) *Saxifragaceae*

Clumping, slowly spreading plant that grows from central point. Leaves shallowly or deeply lobed. In spring, and sometimes again in fall, bare stems carry narrow cones of star-shaped, white or pink flowers. An ever-growing list of cultivars.

CULTIVATION

Shade to part shade and well-drained soil. A good mulch goes a long way; they will take the dry weeks of summer without supplemental water. Cut off spent flower stems, and more flowers may come. Evergreen rosettes can be cleaned up in late winter before flowers emerge.

GARDEN USE

Fabulous shade-garden plants, providing a spark of light in the dark. Regular water needed with more sun exposure. Use with hostas and wild ginger.

SPECIES

T. *cordifolia* 'Oakleaf'—clumping, spreading plant. Blooms midspring to summer. 12 in. x 12 in. Zones 3–7.

CULTIVARS

(All about 12 in. x 12 in; zones 4–9.) 'Black Snowflake' deeply lobed leaves heavily marked with black; pink flowers. 'Iron Butterfly' dark markings along veins; pink flowers. 'Neon Lights' deep lobes marked in black; white flowers. 'Pink Skyrocket' red leaf veins; pink flowers. 'Skeleton Key' dark spot mostly at base of lobes; white flowers.

Tiarella 'Iron Butterfly'

Tradescantia 'Pauline'

Tradescantia

(spiderwort) *Commelinaceae*

Odd bundle of foliage and spiky, angled stems with flowers appearing at the joints, as if they were glued on.

CULTIVATION

Sun or light shade and moist, well-drained soil. Zones 5–9.

GARDEN USE

Blooms early summer to early fall. Plant with other sunny border perennials such as veronicas and mallow. 24 in. x 24 in.

CULTIVARS

Most come from what is called the Andersoniana group: 'Osprey' white flowers with blue eye. 'Pauline' lilac-pink. 'Purple Dome' rose-purple flowers.

Tricyrtis

(toad lily) *Liliaceae*

Small, often spotted, orchidlike flowers at ends of unbranched stems. Upright to arching stems with oblong leaves.

CULTIVATION

Shade to part shade and well-drained, humusy soil. Zones 5–9.

GARDEN USE

Blooms late summer or early fall. Grow with lily-of-the-valley (*Convallaria majalis*), and lungwort (*Pulmonaria*).

SPECIES AND CULTIVARS

T. formosana—leaves clasp stem; spotted flowers. 32 in. x 18 in. 'Gilt Edge' gold striations in leaves; magenta flowers. 'Samurai' creamy white center to leaves.

T. hirta—32 in. x 24 in. 'Miyazaki' white with purple spots. 'Variegata' narrow white margin.

CULTIVARS

'Empress' white with purple spots. 'Tojen' lavender with yellow throat.

Tricyrtis 'Tojen'

Trillium chloropetalum

Trillium

(wake-robin) *Liliaceae*

Three leaves, three petals—everything comes in threes on this woodland plant. Delightful plants of spring, some with flowers that change color as they age.

CULTIVATION

Shade and well-drained, humusy, slightly acidic soil. Grows from rhizomes and does not like being disturbed. Give it room to expand.

GARDEN USE

Grow with like-minded woodland plants.

NOTES

Native habitat of some trilliums has been ravaged by poachers; be sure to buy yours from a nursery you trust.

SPECIES

T. chloropetalum—Northwest native with stemless leaves; mottled purple, pink, to white upward-facing flowers. 16 in. x 8 in. Zones 6–9.

T. erectum—purple-red flowers. 20 in. x 12 in. Zones 4–9.

T. grandiflorum—wavy leaf margins; 3-in.-wide white flowers fade to pink. Easiest to grow (our native). 18 in. x 18 in. Zones 5–8.

T. ovatum—Northwest native similar to T. grandiflorum. 20 in. x 12 in. Zones 5–8.

T. sessile—leaves mottled bronze; brownish red flowers. 12 in. x 12 in. Zones 4–8.

Trollius x *cultorum* 'Lemon Queen'

Trollius

(globe flower) *Ranunculaceae*

Rounded, cup-shaped flowers in bright shades of orange and yellow. Divided or lobed leaves.

CULTIVATION

Sun to part shade in evenly moist soil. Zones 5–8.

GARDEN USE

Grow with mourning widow geranium (*G. phaeum*) in part shade, or veronicas and asters in sun.

SPECIES AND CULTIVARS

T. chinensis—basal, cut leaves; cupped orange flowers. Blooms early summer. 36 in. x 18 in. 'Golden Queen' to 4 ft. x 2 ft.

T. x *cultorum*—dark green, divided foliage; round yellow flowers. Blooms midspring to early summer. 36 in. x 18 in. 'Alabaster' pale cream. 'Lemon Queen' lemon yellow.

Vancouveria

(inside-out flower) *Berberidaceae*

Creeping Western native ground cover. In spring,
sprays of white flowers similar to Epimedium.
Blooms late spring. V. hexandra dies back in winter;
V. planipetala remains evergreen.

CULTIVATION

Shade and well-drained, humusy soil. Zones 5–8.

GARDEN USE

Associates well with wild ginger and epimediums.
Will grow up against, but not up, tree trunks. 12 in. x
12 in.

Vancouveria planipetala

Verbascum

(mullein) *Scrophulariaceae*

Mostly regarded as a biennial, but Verbascum does more than grow one year and flower and die the next, so now mostly termed a "short-lived perennial." Whether tall or short, has a basal rosette of gray or gray-green leaves and one or more spikes of open flowers about an inch across.

CULTIVATION

Full sun and well-drained soil. Zones 4–8.

GARDEN USE

Blooms summer. Add height to the garden by planting tall *Verbascum* at back of borders, but keep shorter selections nearby where you'll enjoy their flowers over a long period.

SPECIES

V. bombyciferum 'Polarsommer'/'Arctic Summer' (giant silver mullein)—big, bold, silver, felted leaves; tall spires of yellow flowers; yes, it's the roadside weed. 8 ft. x 2 ft.

V. chaixii—gray-green rosette of leaves; yellow flowers with fuzzy purple stamen on a tall stalk. 36 in. x 18 in. 'Album' white flowers set off fuzzy stamen. 'Sixteen Candles' yellow flowers; many branched.

V. phoeniceum 'Violetta'—rosette of green leaves; bright purple flowers. Blooms late spring. 36 in. x 18 in.

CULTIVARS

(To 24 in. x 12 in.)

'Caribbean Crush' dark buds, deep coral flowers. 'Helen Johnson' coral. 'Jackie' salmon pink with dark center.

Verbascum 'Helen Johnson'

Verbena bonariensis

Verbena bonariensis

(verbena) *Verbenaceae*

Although most verbenas are annuals in our climate, this is perennial. Tall and narrow; almost leafless, sturdy ridged stems topped with small domes of violet flowers. Blooms all summer.

CULTIVATION

Full sun and well-drained soil. Takes the dry weeks of summer well. Zones 7–11.

GARDEN USE

We are taught to put tall plants in back of borders, but this is a see-through plant that can go any place. Grow with lavender, hebes, and yarrow. Reseeds. 6 ft. x 1.5 ft.

Veronica

(speedwell) *Scrophulariaceae*

A wide range of plants from tall to small. Small, saucer-shaped flowers. Tall veronicas often unbranched. Hebes used to be in the Veronica genus, and you can see the similarities in the spikes of flowers. Leaves are lance-shaped to small and oval.

CULTIVATION

Full sun and well-drained soil. Cut back finished flower spikes of tall veronicas and you'll get more, although smaller, spikes of flowers.

GARDEN USE

There's a veronica for every occasion, from ground cover to spiky clumper. Tall veronicas make good cut flowers.

SPECIES

V. austriaca subsp. *teucrium* 'Crater Lake Blue'—gray-green, slightly hairy foliage; spikes of violet-blue flowers. Blooms all summer. 18 in. x 24 in. Zones 6–8.

V. gentianoides—mat-forming; leafy stems; spikes of pale blue flowers. Blooms early summer. 18 in. x 18 in. Zones 4–8. 'Variegata' white leaf margins.

V. peduncularis 'Georgia Blue'—soft, mat-forming ground cover; leaves turn bronze in cool weather; small blue flowers with white eye. Blooms spring to summer. Sun to part shade. 4 in. x 24 in. Zones 6–8.

V. prostrata—mat-forming; blue flowers on short spikes. Blooms early summer. 6 in. x 16 in. Zones 5–8. 'Heavenly Blue' intense blue. 'Trehane' yellow leaves; sky blue flowers. 2 in. x 12 in.

V. spicata—glossy green, lance-shaped leaves with pointed ends; spikes of small flowers. Blooms early to late summer. 24 in. x 18 in. Zones 3–8. 'Icicle' white. 'Sunny Border Blue' blue.

CULTIVAR

'Giles van Hees' deep pink flowers midsummer to fall. 10 in. x 12 in.

Veronica prostrata 'Trehane'

Viola

(violet) *Violaceae*

We sometimes consider pansies perennials, because they will live over to the next year, but some pansy relatives really are perennial. Heart-shaped leaves. Round petals often marked in center.

CULTIVATION

Part shade and well-drained soil. Zones 5–8.

GARDEN USE

Keep close for a good view—at the edge of the path or in pots.

SPECIES AND CULTIVARS

V. odorata (sweet violet)—egg-shaped leaves; small violet flowers. Blooms late winter to early spring. Stoloniferous. 8 in. x 12 in.

V. riviniana Purpurea group (Labrador violet)—foliage flushed dark; purple flowers. Blooms spring to summer. Spreading, to annoyance of some gardeners, but you can't beat its ground-cover abilities. 8 in. x 16 in.

V. x *wittrockiana* (pansy)—huge collection of cultivars and seed strains that happily reseed in your garden.

Viola x *wittrockiana* hybrid

LIST OF COMMON NAMES

A

Adriatic bellflower (*Campanula garganica*)

Alaska fern (*Polystichum*)

alpine lady's mantle (*Alchemilla alpina*)

alumroot (*Heuchera*)

American burnet (*Sanguisorba canadensis*)

angel's fishing rod (*Dierama*)

anise hyssop (*Agastache foeniculum*)

archangel (*Angelica*)

aster (*Aster*); aster, Stoke's (*Stokesia laevis*)

astilbe

autumn fern (*Dryopteris erythrosora*)

autumn sage (*Salvia greggii*)

avens (*Geum*)

B

baby's breath (*Gypsophila*)

bachelor's button (*Centaurea montana*)

baneberry (*Actaea*)

barrenwort (*Epimedium*)

bats-in-the belfry (*Campanula trachelium*)

beardtongue (*Penstemon*)

bear's breeches (*Acanthus*)

beebalm (*Monarda*)

begonia, hardy (*Begonia grandis*)

bellflower (*Campanula*)

bergamot (*Monarda*)

bergenia (see also elephant ears, pig squeak)

Bethlehem sage (*Pulmonaria saccharata*)

betony (*Stachys*)

bigroot geranium (*Geranium macrorrhizum*)

bishop's hat (*Epimedium*)

bistort (*Persicaria amplexicaule*)

bitterroot (*Lewisia*)

black cohosh (*Actaea racemosa*)

black-eyed Susan (*Rudbeckia*)

blanket flower (*Gaillardia*)

bleeding heart (*Dicentra*)

blue-eyed grass (*Sisyrinchium*)

blue oat grass (*Helictotrichon sempervirens*)

blue star (*Amsonia*)

bouncing bet (*Saponaria officinalis*)

bronze sedge (*Carex buchananii*)

bugbane (*Actaea*)

bugloss, Siberian (*Brunnera*)

burnet (*Sanguisorba*)

bush monkey flower (*Mimulus aurantiacus*)

butterfly weed (*Asclepias*)

natives – columbine

C

California poppy (*Romneya coulteri*)

campanula, see also bellflower, harebell

campion (*Silene*); campion, rose (*Lychnis*)

cardinal flower (*Lobelia*)

carnation (*Dianthus*)

Carpathian bellflower (*Campanula carpatica*)

catchfly (*Silene*)

catmint (*Nepeta*)

checkerbloom (*Sidalcea*)

cheddar pink (*Dianthus*)

Christmas rose (*Helleborus*)

cinquefoil (*Potentilla*)

clematis

clustered bellflower (*Campanula glomerata*)

columbine (*Aquilegia*)

comfrey (*Symphytum*)

coneflower (*Echinacea, Rudbeckia*)

coral bells (*Heuchera*)

coreopsis, see also tickseed

corydalis, see also fumewort, fumitory

cottage pink (*Dianthus*)

cranesbill (*Geranium*)

crimson flag (*Schizostylus coccinea*)

creeping jenny (*Lysimachia*)

creeping phlox (*Phlox stolonifera*)

crocosmia, see also montbretia

cupid's dart (*Catananche caerulea*)

D

daisy (*Leucanthemum* x *superbum*)

Dalmatian bellflower (*Campanula portenschlagiana*)

Dalmatian iris (*Iris pallida*)

dame's rocket (*Hesperis matronalis*)

daylily (*Hemerocallis*)

delphinium

Dittany of Crete (*Origanum dictamnus*)

dropwort (*Filipendula vulgaris*)

E

elephant ears (*Bergenia*)

English primrose (*Primula vulgaris*)

epimedium, see also barrenwort, bishop's hat

evening primrose (*Oenothera*)

F

fairy bells (*Disporum*)

fairy's thimble (*Campanula cochlearifolia*)

false indigo (*Baptisia*)

false mallow (*Sidalcea*)

false Solomon's seal (*Smilacina*)

false sunflower (*Heliopsis*)

feather reed grass (*Calamagrostis* x *acutiflora*)

fern: Alaska (*Polystichum*); autumn (*Dryopteris erythrosora*); Japanese painted (*Athyrium nipponicum*); sword (*Polystichum*); tassel (*Polystichum*)

feverfew (*Tanacetum parthenium*)

flat sea holly (*Eryngium planum*)

flax (*Linum*); flax, New Zealand (*Phormium*)

fleabane (*Erigeron*)

fleeceflower (*Persicaria*)

foamflower (*Tiarella*)

forget-me-not (*Myosotis sylvatica*)

foxglove (*Digitalis*)

fringed bleeding heart (*Dicentra eximia*)

fumewort (*Corydalis*)

fumitory (*Corydalis*)

funkia (*Hosta*)

G

gaura (*Gaura lindheimeri*)

gayfeather (*Liatris spicata*)

gentian (*Gentiana*)

gentian sage (*Salvia patens*)

giant bellflower (*Campanula latifolia*)

giant scabiosa (*Cephalaria gigantea*)

giant sea kale (*Crambe cordifolia*)

giant silver mullein (*Verbascum bombyciferum*)

ginger, wild (*Asarum caudatum*)

gladwyn iris (*Iris foetidissima*)

globe flower (*Trollius*)

globe thistle (*Echinops*)

goatsbeard (*Aruncus*)

golden-eyed grass (*Sisyrinchium californicum*)

golden garlic (*Allium moly*)

golden marguerite (*Anthemis tinctoria*)

goldenrod (*Solidago*)

gooseneck loosestrife (*Lysimachia clethroides*)

H

hardy begonia (*Begonia grandis*)

harebell (*Campanula*)

hellebore (*Helleborus*)

heron's bill (*Erodium reichardii*)

Himalayan blue poppy (*Meconopsis betonicifolia*)

hosta, see also funkia; lily, plantain

Iceland poppy (*Papaver nudicaule*)

I

impatiens

indigo, false (*Baptisia*)

iris

J

Jacob's ladder (*Polemonium*)

Japanese burnet (*Sanguisorba obtusa*)

Japanese painted fern (*Athyrium niponicum*)

joe-pye weed (*Eupatorium*)

Jupiter's beard (*Centranthus ruber*)

K

kale, sea (*Crambe*)

knautia

knotweed (*Persicaria*)

Korean bellflower (*Campanula takesimana*)

lady's mantle (*Alchemilla*)

lamb's ears (*Stachys*)

Lenten rose (*Helleborus*)

ligularia

lily: Peruvian (*Alstroemeria*); plantain (*Hosta*); toad (*Tricyrtis*); torch (*Kniphofia*)

lily-of-the-Nile (*Agapanthus*)

lily-of-the-valley (*Convallaria majalis*)

loosestrife (*Lysimachia*)

lungwort (*Pulmonaria*)

lupine (*Lupinus*)

M

mallow, false (*Sidalcea*)

Maltese cross (*Lychnis*)

marguerite, golden (*Anthemis*)

masterwort (*Astrantia*)

Matijila poppy (*Romneya coulteri*)

meadow rue (*Thalictrum*)

meadow sweet (*Filipendula*)

Michelmas daisy (*Aster novae-belgii*)

milkweed (*Asclepias*)

milky bellflower (*Campanula lactiflora*)

mint hyssop (*Agastache*)

molly the witch (*Paeonia mlokosewitschii*)

monkey flower (*Mimulus*)

monkshood (*Aconitum*)

montbretia (*Crocosmia*)

moor grass, variegated (*Molinia caerulea* 'Variegata')

Moroccan sea holly (*Eryngium varifolium*)

moss pink (*Phlox subulata*)

mother of thousands (*Saxifraga stolonifera*)

mourning widow (*Geranium phaeum*)

Mrs. Robb's spurge (*Euphorbia amygdaloides* var. *robbiae*)

mullein (*Verbascum*)

myrtle spurge (*Euphorbia myrsinites*)

N

navelwort (*Omphalodes*)

nectaroscordum

nettle-leaved bellflower (*Campanula trachelium*)

New England aster (*Aster novae-angliae*)

New Zealand flax (*Phormium*)

New Zealand hair sedge (*Carex comans*)

O

old man's whiskers (*Geum triflorum*)

onion, ornamental (*Allium*)

orange sedge (*Carex testacea*)

oregano, ornamental (*Origanum*)

Oriental poppy (*Papaver orientale*)

ornamental onion (*Allium*)

ornamental oregano (*Origanum*)

P

pasque flower (*Pulsatilla vulgaris*)

peach-leaved bellflower (*Campanula persicifolia*)

penstemon, see also beardtongue

peony (*Paeonia*)

Peruvian lily (*Alstroemeria*)

phlox

pig squeak (*Bergenia*)

pincushion flower (*Scabiosa*)

pink (*Dianthus*)

pink cow parsley (*Chaerophyllum hirsutum* 'Roseum')

plantain lily (*Hosta*)

poker plant (*Kniphofia*)

poppy (*Meconopsis, Papaver*); poppy: California (*Romneya coulteri*)

poppy-flowered anemone (*Anemone coronaria*)

potentilla, see also cinquefoil

prairie smoke (*Geum triflorum*)

primrose (*Primula*); primrose, evening (*Oenothera*)

purple coneflower (*Echinacea purpurea*)

pyrethrum (*Tanacetum*)

Q

queen of the prairie (*Filipendula*)

R

ragged robin (*Lychnis*)

rattlesnake master (*Eryngium yuccifolium*)

red hot poker (*Kniphofia*)

rock soapwort (*Saponaria ocymoides*)

rodgersia

rose campion (*Lychnis*)

rue (*Ruta graveolens*)

Russian sage (*Perovskia atriplicifolia*)

rusty foxglove (*Digitalis ferruginea*)

S

sage (*Salvia*)

salvia, see also sage

sandwort (*Arenaria*)

saxifrage (*Saxifraga*)

scabiosa, giant (*Cephalaria gigantea*)

scarlet hedge nettle (*Stachys coccinea*)

scarlet monkey flower (*Mimulus cardinalis*)

sea campion (*Silene uniflora*)

sea holly (*Eryngium*)

sea kale (*Crambe*)

sedge (*Carex*)

sedum, see also stonecrop

Serbian bellflower (*Campanula poscharskyana*)

Shasta daisy (*Leucanthemum* x *superbum*)

Siberian bugloss (*Brunnera*)

Siberian iris (*Iris sibirica* x *Iris sanguinea*)

silver grass (*Miscanthus sinensis*)

snakeroot (*Actaea*)

sneezeweed (*Helenium autumnale*)

soapwort (*Saponaria*)

soldiers-and-sailors (*Pulmonaria*)

Solomon's seal (*Polygonatum*); Solomon's seal, false (*Smilacina*)

speedwell (*Veronica*)

spiderwort (*Tradescantia*)

spotted bellflower (*Campanula punctata*)

spotted geranium (*Geranium maculatum*)

spurge (*Euphorbia*)

star of Persia (*Allium cristophii*)

stinking hellebore (*Helleborus foetidus*)

stinking iris (*Iris foetidissima*)

Stoke's aster (*Stokesia laevis*)

stonecrop (*Sedum*)

strawberry foxglove (*Digitalis* x *mertonensis*)

strawberry saxifrage (*Saxifraga stolonifera*)

sundrops (*Oenothera*)

sunflower (*Helianthus*); sunflower, false (*Heliopsis*)

swamp sunflower (*Helianthus angustifolius*)

sweet violet (*Viola odorata*)

sword fern (*Polystichum*)

T

tansy (*Tanacetum*)

tassel fern (*Polystichum*)

Texas betony (*Stachys coccinea*)

threadleaf coreopsis (*Coreopsis verticillata*)

thrift (*Armeria*)

tickseed (*Coreopsis*)

toadflax, purple (*Linaria purpurea*)

toad lily (*Tricyrtis*)

torch lily (*Kniphofia*)

trillium, see also wake-robin

V

variegated moorgrass (*Molina caerulea* 'Variegata')

verbascum, see also mullein

veronica, see also speedwell

violet (*Viola*)

W

wake-robin (*Trillium*)

wallflower (*Erysimum*)

water avens (*Geum rivale*)

Welsh poppy (*Meconopsis cambrica*)

white globe thistle (*Echinops sphaerocephalus*)

wild blue phlox (*Phlox divaricata*)

wild ginger (*Asarum caudatum*)

willow blue star (*Amsonia tabernaemontana*)

willow gentian (*Gentiana asclepiadea*)

willowleaf sunflower (*Helianthus salicifolius*)

windflower (*Anemone*)

wine loosestrife (*Lysimachia atro-purpurea* 'Beaujolais')

wood anemone (*Anemone nemorosa*)

wood aster (*Aster divaricatus*)

wood spurge (*Euphorbia amygdaloides*)

wormwood (*Artemesia*)

Y

yarrow (*Achillea*)

yellow coneflower (*Echinacea paradoxa*)

yellow foxglove (*Digitalis grandiflora*)

yellow stork's bill (*Erodium chrysanthum*)

INDEX

Note: Photographs are indicated by
italics.

prairie smoke, *158*, 159

primrose (*Primula*), 29, 234–35, *234*

primrose, evening (*Oenothera*), *210*, 211

Primula, 29, 234–35, *234*

Prosartes, 131, *131*

Pulmonaria, 7, 31, 40, *236*, 237

Pulsatilla vulgaris, 238, *238*

purchasing plants, tips for, 8–10, 12–13, 39

purple coneflower, *14*, *133*, 134

purple toadflax, *8*, 24, 38, 190, *190*

pyrethrum, 268, *269*

Q

queen of the prairie, 146–47, *147*

R

ragged robin, 195–96, *195*

rattlesnake master, 141

red hot poker, 183–84, *183*

repetition, 24

replanting seedlings, 46

reseeding perennials, 32–33, 38

Rhododendron schlippenbachii, 31

rock soapwort, 29, *46*, 250, *250*

Rodgersia (rodgersia), 23, 239–40, *239*

Romneya coulteri, *240–41*, 241

roots, 12, 18–19, 47, 48

rose, Christmas, vi, 169

rose, Lenten, 169

rose campion, 44, 195–96, *195*

roses, 31

royal azalea, 31

Rudbeckia, 242, *243*

rue (*Ruta graveolens*), 244–45, 245

rue, meadow (*Thalictrum*), 33, *270*, 271

Russian sage, 220–21, 221

rusty foxglove, *129*, 130

Ruta graveolens, 244–45, 245

S

sage (*Salvia*), 246–47, *246*

sage, autumn, *246*, 247

sage, Bethlehem, 237

sage, gentian, 247

sage, Russian, *220–21*, 221

sales, plant, 10, 39

Salvia, 246–47, *246*

Sanguisorba, 248, *248–49*

Saponaria, *46*, 250, *250*

    See also soapwort

Saxifraga, 251, *251*

saxifrage, 251, *251*

Scabiosa (pincushion flower), 37, *42*, 252, *253*

scabiosa, giant (*Cephalaria gigantea*), *23*, 112, *112*

scale, 26

scarlet hedge nettle, 263

scarlet monkey flower, 200

Schizostylus coccinea, 29, 254, *254*

scientific names, 4–5

sea campion, 258, *258*

sea holly, *140*, 141

sea kale, 120, *120*

seasons

    buying tips, 10, 12

    choosing plants for, vi, 3, 6–8

    container planting, 16, 17, 40, 47

    garden design, 6–8, 24, 28

    hardiness zones, 7

    maintenance tasks, 32–33, 36, 38, 39–49

    planting tips, 15–16

Native – Sedum

FEATURED GARDENS

I would to thank the many people who invited me into their gardens and made this book possible: Swanson's Nursery, Seattle; Wells Medina Nursery, Bellevue, p 30; Rosehip Farm and Garden, Coupeville, Whidbey Island; Bayview Farm and Garden, Langley, Whidbey Island; Center for Urban Horticulture at the University of Washington, Seattle, WA; The Elisabeth C. Miller Botanical Garden, Seattle, WA, p 10, 45; The Northwest Perennial Alliance border at the Bellevue Botanical Garden, Bellevue, WA; The Johnson and Beebe Garden, Whidbey Island, WA; the Crooks Garden, Seattle, WA, p 18, 31, 317; the Emerson Garden, Whidbey Island, WA; the Endsley Garden, Seattle, WA, p 5; the Hand Garden, Seattle, WA; the Henry Garden, p 42; the Hulbert Garden, Seattle, WA, p 46; the Juntunen Garden, Mount Vernon, WA, p 44; the Stapp Garden, Seattle, WA; the Dohna Garden, Vashon Island, WA, p 24; the Page Garden, Portland, OR, p 13; the Roberson Garden, p 28; the Smith Garden, Seattle, WA, p 32; the Tong Garden; Seattle, WA, p 21; the Newton Garden, Vancouver, BC, Canada, p 14. —JK

ACKNOWLEDGMENTS

Compiling a comprehensive list of perennials for Northwest gardens is a lot like herding cats. Besides the obvious qualifications—what makes a good garden plant?—are all the vagaries of horticulture. Cultivars land big on the scene one year and have disappeared two years later. Species are shifted from one genus to another. Nursery shelves are flooded with a particular selection while the plant is non-existent in catalogs (or vice versa).

That may sound as if compiling information and writing up descriptions of perennials was a hardship—far from it. My own garden swells and shrinks with new plants (and a few losses) when I carry out hands-on research. It's an ongoing and satisfying process, and I owe thanks to all the gardeners and nursery workers who have advised me through the years.

ABOUT THE AUTHOR

Marty Wingate writes and speaks about gardens and travel. She is the author of three previous books on gardening, including *Landscaping for Privacy*. She is also the author of the Potting Shed mystery series about an American gardener in England.

Marty's garden articles appear in a variety of publications, and she is a weekly guest on KUOW (94.9 FM), Seattle's NPR station. Marty leads garden tours to European and North American destinations. Her website is: MartyWingate.com.

ABOUT THE PHOTOGRAPHER

Jacqueline Koch has photographed many gardens in the Pacific Northwest and is currently the director of communications for J/P Haitian Relief Organization. She divides her time between Seattle, WA, and Port-au-Prince, Haiti.